Lecture Notes in Logistics

More information about this series at http://www.springer.com/series/11220

Ana Paula Barbosa-Póvoa
Albert Corominas · João Luís de Miranda
Editors

Optimization and Decision Support Systems for Supply Chains

Springer

Editors
Ana Paula Barbosa-Póvoa
Department of Engineering and Management
Instituto Superior Técnico
Lisbon
Portugal

Albert Corominas
Department of Management
Institute of Industrial and Control Engineering
Barcelona
Spain

João Luís de Miranda
ESTG/IPP
Portalegre
Portugal

and

CERENA/IST
Lisboa
Portugal

ISSN 2194-8917 ISSN 2194-8925 (electronic)
Lecture Notes in Logistics
ISBN 978-3-319-42419-4 ISBN 978-3-319-42421-7 (eBook)
DOI 10.1007/978-3-319-42421-7

Library of Congress Control Number: 2016945779

Printed on acid-free paper

This Springer imprint is published by Springer Nature
The registered company is Springer International Publishing AG Switzerland

Preface

Since their inception in 1982, supply chain and supply chain management (SC/SCM) has come to achieve a dominant role as integrative concepts concerning the broad field of management. Among the reasons for this success can be pointed out, first, that SC/SCM is an updated attempt to give a systemic view of activities and flows that occur when supplying, producing, distributing and recovering one or more products and, second, that they establish the appropriate framework to describe, analyse and solve the new problems that have emerged as a consequence of the evolution of the increasingly complex real production systems. Hence, powerful decision support system (DSS) and optimization tools are required to deal with the new management challenges.

The triennial Erasmus project "Optimization and Decision Support Systems for Supply Chains" that was held in the College of Technology and Management, Portalegre Polytechnics Institute, Portalegre, Portugal, from 2011 to 2014, was conceived, in the light of the preceding considerations, as a contribution to help researchers and practitioners to face the problems that arouse in supply chains. It has been for us a pleasure to edit and now introduce the present volume as a follow-up of the project.

The Erasmus Intensive Programme (IP) explored the training on modelling and optimization of production–distribution facilitiy networks, considering material and financial flows in a multi-echelon system, while also addressing a green logistic approach to supply chains.

The contributions for this volume are mainly based on the IP lectures, which addressed transversal and complementary SC topics: from the main topics on SCM until optimization advanced techniques, and by covering either information systems to planning and scheduling of production processes usually found in manufacturing or petrochemical. In addition, SC sustainability and reverse logistics have been also treated, being the computational sessions supported by IBM.

During three academic years, more than one hundred participants originated from more than twenty countries got together to share the most recent advancements on industry-based SC and to discuss their current challenges. Thus, as a result and in line with the IP purposes, this book is developed and directed for M.Sc./Ph.D. students or researchers on engineering/logistics or mathematical specialties and industry professionals.

Thus, this book on "Optimization and Decision Support Systems for Supply Chains" hopefully will serve as an important reference to the European higher education area on SC, namely providing an overview of very important SC topics that can be used for M.Sc./Ph.D. works and helping SC researchers and practitioners in practical developments.

Finally, we would like to thank all the contributors for their quality manuscripts, so as to the reviewers for their due time appreciations.

Lisbon, Portugal Ana Paula Barbosa-Póvoa
Barcelona, Spain Albert Corominas
Portalegre, Portugal João Luís de Miranda

Acknowledgements

Lecturers from Partner Institutions

Professor Ana Cristina Amaro ISCAC—Coimbra Business School, Instituto Politécnico de Coimbra

Professor Giuseppe Bruno Dipartimento di Ingegneria Economico-Gestionale, Università degli Studi di Napoli "Federico II"

Dr. Andrea Genovese Management School, University of Sheffield

Professor João Leitão CEG-IST, Instituto Superior Técnico, Universidade de Lisboa; UBI, Universidade da Beira Interior

Dr. Samuel Moniz CEG-IST, Instituto Superior Técnico, Universidade de Lisboa; INESC-TEC, Universidade do Porto

Talkers from Collaborating Companies

Dr. Rui Barrento Hutchinson Group, Portalegre site

Dr. Raquel Carpinteiro IMG Group, Evertis/Selenis, Portalegre site

Dr. Nuno Dias Jeronimo Martins Group

Dr. Lukasz Milanowski Hutchinson Group, Portalegre site

Dr. Sandra Nicolau Valnor, Avis/Fronteira site

Dr. Carlos Paiva IMG Group, Evertis/Selenis, Portalegre site

Dr. Claudia Simões Valnor, Avis/Fronteira site

Support

Partners

UNIVERSITAT POLITÈCNICA
DE CATALUNYA
BARCELONATECH

GÖTEBORGS UNIVERSITET

The
University
Of
Sheffield.

Contents

Contributors

Ana Paula Barbosa-Póvoa Centro de Estudos de Gestão, Instituto Superior Técnico, Lisbon University, Lisbon, Portugal

Ana Carvalho CEG-IST, Instituto Superior Técnico, Universidade de Lisboa, Lisboa, Portugal

Miguel Casquilho Department of Chemical Engineering, Instituto Superior Técnico, Lisbon, Portugal; CERENA—Centro de Recursos Naturais e Ambiente, Instituto Superior Técnico, University of Lisbon, Lisbon, Portugal

Dulce Coelho Polytechnic Institute of Coimbra - ISEC, Rua Pedro Nunes, Coimbra, Portugal; INESC Coimbra, Coimbra, Portugal

Juan Manuel Garcia-López IBM—ILOG Decision Optimization, Madrid, Spain

Jorge Gonçalves PETROGAL SA, Lisbon, Portugal

Kseniia Ilchenko Kyiv Polytechnic Institute, National Technical University of Ukraine, Kyiv, Ukraine

Beatriz Jiménez-Parra Department of Management and Business Economics, Universidad de León, León, Spain

Marco Laumanns IBM Research, Rueschlikon, Switzerland

Amaia Lusa Department of Management, Institute of Industrial and Control Engineering, Universitat Politècnica de Catalunya, Barcelona, Spain

Johan Magnusson Department of Applied IT, Chalmers and University of Gothenburg, Gothenburg, Sweden; Westerdal Oslo School of Art, Communication and Technology, Oslo, Norway

Henrique A. Matos CERENA, Instituto Superior Técnico, Universidade de Lisboa, Lisboa, Portugal

João Luís de Miranda Departamento de Tecnologias e Design, Escola Superior de Tecnologia e Gestão, Instituto Politécnico de Portalegre, Portalegre, Portugal; CERENA—Centro de Recursos Naturais e Ambiente, Instituto Superior Técnico, University of Lisbon, Lisbon, Portugal

Nuno Mota Centro de Estudos de Gestão CEG-IST, Instituto Superior Técnico, Universidade de Lisboa, Lisbon, Portugal

Olga Nazarenko Kyiv Polytechnic Institute, National Technical University of Ukraine, Kyiv, Ukraine

Andreas Nilsson Department of Applied IT, Chalmers and University of Gothenburg, Gothenburg, Sweden

Carla Oliveira Henriques Polytechnic Institute of Coimbra - ISCAC Business School, Coimbra, Portugal; INESC Coimbra, Coimbra, Portugal

Jordi Olivella Nadal Department of Management, Institute of Industrial and Control Engineering, Universitat Politècnica de Catalunya, Barcelona, Spain

Susana Relvas Centro de Estudos de Gestão CEG-IST, Instituto Superior Técnico, Universidade de Lisboa, Lisbon, Portugal

Sergio Rubio Department of Business Management & Sociology, Universidad de Extremadura, Badajoz, Spain

Stefan Woerner IBM Research, Rueschlikon, Switzerland

Optimising Sustainable Supply Chains: A Summarised View of Current and Future Perspectives

Ana Paula Barbosa-Póvoa

Abstract Sustainable supply chain (SSC) research is recognised as being an emerging field where both academics and industry communities have an important role to play. At the organizations level, companies should not only be managed towards a profitability goal but also their contribution towards a sustainable society needs to be part of their decisions. For academic's sustainable supply chains although being an area that has been the focus of several studies is still far from being consensual. On the optimization of supply chains different perspectives have been adopted but much further has to be done. Along this chapter the concept of sustainable supply chains and some of the important publications in the area are analysed. A particular focus is given to the construction of tools based on optimization that may help the decision process in such systems while improving the efficiency and responsiveness of SSC. The chapter concludes with a recap of the published work while identifying key aspects that should be pursued so as to build optimized sustainable supply chains.

Keywords Sustainability · Sustainable supply chain · Optimization · Closed-loop supply chains · Design · Planning

1 Introduction

Supply chains activities have evolved over time and similarly has their definition. Now a days some convergence exists on how to define such systems. As stated by Christopher (2012) a supply chain can be defined as a network of organizations that are involved, through upstream and downstream linkages, in the different processes and activities that produce value in the form of products or services delivered to the ultimate consumer. Sustainability in such systems has been a recent focus but the

---

A.P. Barbosa-Póvoa (✉)
Centro de Estudos de Gestão, Instituto Superior Técnico, Lisbon University, Lisbon, Portugal
e-mail: apovoa@tecnico.ulisboa.pt

A.P. Barbosa-Póvoa et al. (eds.), *Optimization and Decision Support Systems for Supply Chains*, Lecture Notes in Logistics,
DOI 10.1007/978-3-319-42421-7_1

concept of sustainable supply chains is still undefined and some misperception of it exists. While sustainable development describes a global view of people and nature, efforts to achieve this in practice provoke questions on how this concept could be meaningfully interpreted and operationalized for smaller systems, such as supply chains. A close-systems view tries to apply the concept of sustainability just to a specific enterprise itself, while an open-systems view focuses on how the organization contributes to sustainable development of a wider society (Figge and Han 2004). Both perspectives have their inherent strengths and neither one should be readily discarded. This must be reflected in the way the sustainability concept is applied at the supply chain level. Successful sustainability-driven supply chain should contribute positively to the sustainable development of the larger social-ecological system of which it is part (Parrish 2009).

The first step towards understanding how sustainable supply chains are shaped and how they evolve over time implies getting more deeply involved with the phenomenon called "sustainability". Using the definition of the World Commission on Environment and Development, sustainability can be viewed as the use of resources to meet needs of the present without comprising the ability of future generations to meet their own needs (WCED 1987). Thus, sustainability definitely influences organizations and it certainly influences the shape of current and future supply chains. Combining this with the fact that production is increasingly fragmented across geographic spaces and between companies, the need to develop a more profound comprehension of the interaction between these elements within the supply chain is demanding. In particular, the need to seriously explore the concept of sustainable supply chains within a collaborative perspective should be further explored and taken as a goal of any organization. This would help companies' revenue growth, costumer's recognition and their contribution to society and to the planet (Kleindorfer et al. 2005). To respond to this challenge companies must invest on the design, planning and operation of their supply chains while considering the minimization of their global energy consumption (Barbosa-Póvoa 2009). On way of helping such goal is through the usage of optimization tools that will support the supply chains decision process (Grossmann 2004; Suring 2013).

In this chapter the analysis on how the concept of sustainability should be incorporated into supply chains is explored. Then and having in mind that tools are required to help the decision making process of sustainable supply chains that are complex systems a focus is given to the use of optimization to build such tools. Finally the chapter concludes with the identification of future challenges in the area of sustainable supply chains optimization.

2 The Extended Supply Chains Concept

Sustainable Supply Chains Management has its roots in both Sustainability and Supply Chain Management literature and involves a broadened approach of the supply chain management. The concept of sustainable supply chains is now

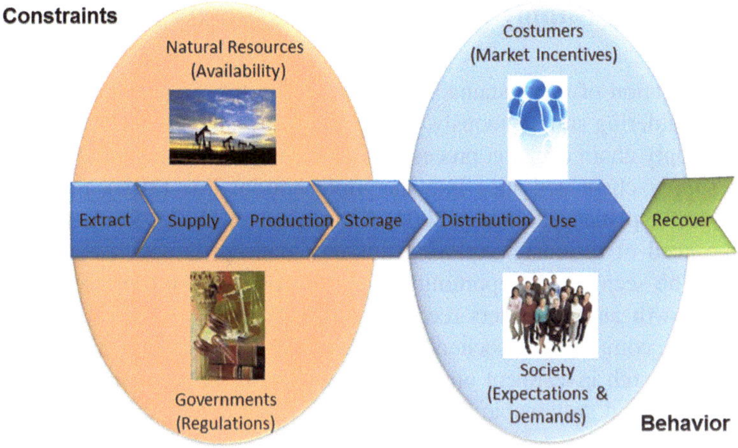

Fig. 1 Extended Supply Chains

emerging (Suering and Muller 2008; Barbosa-Póvoa 2009; Meckenstock et al. 2015) and it is now clear that supply chains need to be defined as operational structures that are able to manage raw materials and services from suppliers to manufacturers/service providers and back while guaranteeing costumers satisfaction and contributing positively to society and to the planet. An extended supply chains view must then be adopted as depicted in Fig. 1. Here an external supply chain view is considered where supply chain activities are integrated and forward and reverse flows are managed not in an isolated form but simultaneously while surrounded by an extended context that includes costumers and marketing incentives, natural resources availability, governments and regulations constraints and society expectations and demand.

Furthermore, and considering an internal supply chains view a new understand should be given to the well-known supply chains activities as stated by Linton et al. (2007). These include product design, production, distribution, and the management of end of life products. Such activities should be managed considering their contribution to the economic goal but also environmental and societal objectives should be assimilated (Barbosa-Póvoa 2009).

The integration between the extended external and internal views of the supply chains is mandatory as the basis to construct sustainable supply chains. Efforts on both sides will support the goal of responsible care as stated by Grossmann (2004), but require support on the decision process. The development of systematic methods and tools that guarantee the design of environmental benign products and processes while creating economic and societal optimized conditions are an end to be accomplished.

3 Optimising Supply Chains Towards Sustainability

The incorporation of the sustainability concept into the supply chain management implies considering simultaneously economic, environmental as well social aspects into the supply chain decision process. This meaning increasing the engagement of all the supply chain activities into the triple bottom line (3BL), the threes P's, people, profit, planet (Elkington 2004; Kleindorfer et al. 2005). At the company level the need of seriously exploring this new concept is now a reality and it is starting to be seen as an opportunity as it can contribute to improve companies' revenue growth and costumers recognition (Barbosa-Póvoa 2009; UNGC 2013). However, the complexity associated with this new supply chains paradigm calls for academic research investment on the development of methods and tools that may help to understand sustainable supply chains fostering companies decision making process support.

Different approaches have been followed by the academics when addressing sustainability in supply chains. Some authors have looked only to the forward supply chains while others have explored the reverse logistics as a way to deal with the environmental and social pressures. The closed-loop supply chain concept has materialized where both flows, forward and reverse, are considered simultaneously. As referred by Guide and Van Wassenhove (2002) the companies that have been most successful with their reverse supply chains are those that closely coordinate them with the forward supply chains, creating the closed-loop supply chain (CLSC). These companies have a positive contribution to the sustainable development by defining and operating the right systems that guarantee a reduction of resources consumption by recycling and recovering end of life products back into the chain. Such systems have been studied since 1999 by academia and in 2007 as stated by Salema et al. (2007) the literature on closed-loop supply chains was already been slowly building. Nowadays and as shown by Cardoso et al. (2013) closed loop supply chains are a reality and some effort has been done by academia on the development of optimization tools.

Jayaraman et al. (1999) developed a mixed integer programming formulation to model a closed-loop supply chain. The model was tested on a set of problems based on the parameters of an existing electronic equipment remanufacture firm. Fleischmann et al. (2000) proposed a model for the location of logistic facilities. Their work was the first to propose a general model formulation (Mixed Integer Linear Programming Model, MILP) that simultaneously optimizes the reverse and the forward networks. Two case studies were used to explore the model application. Krikke et al. (2003) also proposed a MILP model for the design of a closed-loop supply chain where both location-allocation decisions and product design were considered. The objective function included both supply chain costs and environmental impacts using a performance indicator (based on LCA approach). As product design was involved, an assembly and disassembly of products was explicitly modelled. Beamon and Fernandes (2004) presented a model for a single product closed-loop supply chain design. Fandel and Stammen (2004) propose a strategic model for the supply chain design. Salema et al. (2006, 2007, 2009, 2010) studied the problem of designing simultaneously the forward and reverse networks. In such works the models evolved along time and different aspects of the closed-loop supply chains have been considered. The proposed models are fairly general as incorporate facility capacity limits, multi-product flows and uncertainty. Strategic versus tactical decisions were considered. Quarasigui Frota Neto et al. (2008) developed a framework for the design of logistics structures in which profitability and environmental impacts are balanced. Zeballos et al. (2012) introduced a two-stage scenario-based modelling approach in order to deal with the design and planning decisions in multi-period, multi-product CLSCs subject to uncertain conditions. Uncertainty is associated to the quantity and quality of the flow of products of the reverse network.

Ilgin and Gupta (2010) offer a description of the main type of modelling techniques and topics addressed in close-loop supply chain research. Stindt and Sahamie (2014) analyse close-loop supply chain research in different sectors of the process industry.

Dekker et al. (2012) state that most papers focused on close-loop supply chain do not explicitly deal with the supply chain environmental impacts, and draw attention to the need for new models to support environment related decision making.

Recently, Cardoso et al. (2013), proposed a detailed dynamic model for the design and planning of supply chains where reverse logistics activities are considered simultaneously with forward supply chain activities. Supply chain dynamics are incorporated through capacity expansion and dynamic entities links along the planning horizon. Uncertainty in products demands is considered, justified by markets' volatility. Also, Zebalos et al. (2014) addressed uncertainty in CLSC. A Mixed Integer Linear Programming (MILP) multi-stage stochastic model is proposed to deal with the design and planning problem of multi-period multi-product closed-loop supply chains. The MILP formulation is proposed for addressing general CLSCs, structured as a 10-layer network (5 forward plus 5 reverse flows), with uncertain levels in the amount of raw material and customer demands. The effects of uncertain demand and supply on the network are considered by means of scenarios. The goal is to minimize the expected cost of facilities, purchasing, storage, transport and emissions, minus the expected revenue due to returned products. To show the application of the mathematical formulation, several instances of an example proposed in the literature are examined.

The above works focused essentially on the strategic and at most on the planning contexts. At the operational level fewer works appeared that dealt with the closed-loop supply chains. Amaro and Barbosa-Póvoa (2008a) looked into the detailed scheduling of supply chains with reverse flows where different product recovery policies are analysed. A real case-study of the pharmaceutical industry was solved. Later on, the same authors Amaro and Barbosa-Póvoa (2008b) presented a generic approach where the integration between the planning and the scheduling level was proposed. The model also contemplated the reverse flows where the management of non-conform and end-of-life products was considered.

On the reverse logistic structures several works have also been reported. A review, on the characteristics of the research on reverse logistics, is presented by Rubio et al. (2008). Ammons et al. (2000) developed a MILP model for the design and planning of reverse production systems. The State-Task Network (STN) was used as representation methodology. The model was applied to the recycling of a network router from an Original Equipment Manufacturer. The same problem was addressed by Realff et al. (2000) through a different approach where a robust optimization framework was used. Duque et al. (2007), using the maximal State-Task Network proposed an optimization model for the design and operation of a recovery route for residual industrial products with uncertain demands. The final model is able to suggest the optimal processing and transport routes, while optimizing a given objective function and meeting design and environmental constraints. This work was extended in Duque et al. (2009) where the eco-indicator 99 (Pré consultants 2001) was used to quantify the environmental and social impact of the chain. In the same year, Quarisiguai Frota Neto et al. (2009) studied the eco-efficiency methodology and proposed a multi-objective linear problem with three objectives: minimize costs, cumulative energy demand and waste in a reverse logistics network. Gomes et al. (2011) addressed the design and planning of reverse chains and proposed a generic MILP model where the best locations for collection and sorting centres are chosen simultaneously with the definition of a tactical network planning. The model was applied to a real case on the collection of waste electric and electronics equipment (WEEE) products.

On the forward chains the sustainability concept has been mainly associated with the quantification of environmental impacts. Zhou et al. (2000) developed a goal programming model to account for sustainability aspects on the supply chains of continuous processes. Hagelaae et al. (2002) explored the concept of Life Cycle Analysis (LCA) when applied to a supply chain context and concluded that no guidelines exist for an integration of these two strategies. Turkay et al. (2004) studied the problem of multi-company collaborative supply chain management where not only economical goals where considered but also environmental aspects where incorporated. Hugo and Pistikopoulos (2005) looked into the supply chains planning problem where environmental concerns were accounted for. A mathematical formulation was developed which was applied to a bulk chemicals supply chain. In 2006, Soylu et al. (2006) analysed the synergies that may exist on a collaborative supply chain of energy systems. Matos and Hall (2007) analysed the integration of sustainable development concepts into the supply chains and build up a framework that should guide the utilization of LCA methods into the supply chains. Two case studies were studied, oil sand refining and agriculture biotechnology. Bojarski et al. (2009) also studied the design and planning of forward supply chains considering economic and environmental aspects. Guillén-Gosalbez and Grossmann (2010) addressed the design and planning of forward chemical supply chains where environmental concerns are modelled in the presence of uncertainty in the life cycle inventory of the supply chain operation. More recently, Pinto-Varela et al. (2012), presented a MILP formulation for the design of supply chains with reverse flows where environmental aspects were incorporated, and

Amin and Zhang (2012) proposed a MILP to optimize a closed-loop supply chain based on product life cycle.

From above it can be seen that the sustainability concept has been studied but not yet fully addressed in an integrated form. The integration of the tree pillars of sustainability is still to be explored. The social component has not yet been analysed. Hassini et al. (2012) corroborate this conclusion claiming that none of the measures described in their review have been designed to be used in a supply chain context. A recent work by Carvalho and Barbosa-Póvoa (2011) developed an analysis of possible social indicators that can be used to evaluate the social responsibility performance in global supply chains. You et al. (2012) determine the social benefit of a cellulosic biofuel supply chain, measured through full-time equivalent yearly jobs created. Mota et al. (2015) presented a detailed multi-objective optimization methodology that accounts for economic, environmental and social concerns in a supply chain with reverse flows. Environmental impact assessment is considered through the use of ReCIPe 2008 (Goedkoop et al. 2009). A social benefit indicator is developed where the creation of employment in less developed regions is preferred. The multi-objective approach is used to reach a solution of compromise between the three sustainability pillars. The model is applied to a case study developed in collaboration with a Portuguese company, leader in battery production.

The above literature analysis clearly shows that the field on sustainable supply chain optimization is still in its infancy but important ideas have already been explored. However a clear consensus on the analysis of such systems and more detailed forms of looking into the integration of the three sustainability pillars within the supply chains optimization should be further explored.

4 Conclusions and Future Research Agenda

Along this chapter it was shown that an increased concern with sustainability at the supply chain level exists. Sustainable supply chains have become important not only within the academic community but also at the industrial level. Within this context the development of tools that can support the decision process of these emergent complex systems—sustainable supply chains, has been recognised as mandatory. Several works have been published on the topic but it is also clear that much is still to be done.

Sustainability in the supply chains has been addressed in different forms. This range from the explicit consideration of environmental issues into forward supply chains, to the modelling of the simultaneous forward and reverse flows where collection, recycling and remanufacturing of non-conforms and/or end-of-life products, have been well thought-out. Moreover, the simultaneous consideration of closed-loop supply chains and environmental impacts was at the same time considered with the integration of detailed process activities into the supply chain. But further work is still required to fully address the optimization of sustainable supply chains where the different aspects of sustainability become adequately modelled. To achieve such target the research community has to look into a set of key aspects and some answers are to be given:

1. How to measure in an appropriate form sustainability into the supply chains? Further work needs to be done on the environmental impacts measures and its consequences in the social wellbeing of the population. The incorporation of social concerns into the supply chains is still an area unexplored and almost no measures exist to quantify properly this component.
2. How to define or redesign the supply chains to facilitate collection, refurbishment, recycling or disposal or returned products? The optimization of the supply chains structures to easily incorporate the end-of-life products is required. Although, several studies appeared on the closed-loop supply chain, general models are still scarce and its application to a large number of real supply chains has yet to been done.
3. How to trade-off the profit, people and planet of the 3BL into the supply chains management? Multi-objectives approaches need to be explored, which encompass all the above aspects.
4. How to deal with uncertainty? Uncertainty influences the sustainable supply chains both at the supply chain structure definition as well as at the associated planning and scheduling activities. Does a detailed form of dealing with uncertainty is required.
5. How to minimize risk when building and operating sustainable supply chain structures? Different decisions lead to different levels of risk that need to be defined so as to take the right decision. The need of identifying the appropriate risk measures that will help the establishment of proper trade-offs between profit and risk, amongst others objectives is a challenge to attain.
6. How to incorporate responsiveness and resilience into sustainable supply chains? At a qualitative level supply chain risk management strategies dealing with redundancy and flexibility have been proposed in the literature but these strategies have not yet been considered explicitly into the optimization models. The need of quantifying resilience and responsiveness in supply chains requires further investment.

The study of the above points will help to expand the scope and nature of sustainable supply chains optimization and contribute to effectively solve real-world industrial problems.

Concluding, following the above sketched research lines should aim to build holistic models that will support sustainable supply chains management decisions while accounting for the three pillars of sustainability (economic, environmental and social) under an increasing uncertain context.

References

Amaro, A. C., & Barbosa-Póvoa, A. P. F. D. (2008a). Optimal supply chain management with detailed scheduling. *Industrial and Engineering Chemistry Research, 47*(1), 116–132.

Amaro, A. C., & Barbosa-Póvoa, A. P. F. D. (2008b). Planning and scheduling of industrial supply chains with reverse flows: A real pharmaceutical case-study. *Computers & Chemical Engineering, 32*, 2635–6006.

Amin, S., & Zhang, G. (2012). A proposed mathematical model for closed-loop network configuration based on product life cycle. *International Journal of Advanced Manufacturing Technology, 58*, 791–801.

Ammons, J. C., Realff, M., & Newton, D. (2000). *Decision models for reverse production system design* (pp. 341–362). Bonton: Handbook on Environmentally Conscious Manufacturing. Kluwer.

Barbosa-Povoa, A. P. (2009). Sustainable supply chains: Key challenges. 10th international symposium on process systems engineering—PSE2009. *Computer Aided Chemical Engineering, Computer-Aided Chemical Engineering, 27*, 127–132.

Beamon, B. M., & Fernandes, C. (2004). Supply-chain network configuration for product recovery. *Production Planning & Control, 15*, 270–281.

Bojarski, A., Laínez, J., Espuña, A., & Puigjaner, L. (2009). Incorporating environmental impacts and regulations in a holistic supply chains modeling: An LCA approach. *Computers & Chemical Engineering, 33*, 1747–1759.

Cardoso, S., Barbosa-Póvoa, A. P., & Relvas, S. (2013). Design and planning of supply chains with integration of reverse logistics activities under demand uncertainty. *European Journal of Operations Research, 226*(3), 4363.

Carvalho, A., & Barbosa-Póvoa, A. P. F. D. (2011), Social indicators for supply chain analysis, EuroMa 2011, Cambridge.

Christopher, M. (2012). Logistics and supply chain management. UK: Pearson.

Dekker, R., Bloemhof, J., & Mallidis, I. (2012). Operations research for green logistics–An overview of aspects, issues, contributions and challenges. *European Journal of Operational Research, 219*, 671–679.

Duque, J., Barbosa-Póvoa, A. P. F. D., & Novais, A. Q. (2007). Synthesis and optimisation of the recovery route for residual products with uncertain product demands. *Computers & Operations Research, 34*, 1463–1490.

Duque, J., Barbosa-Póvoa, A. P. F. D., & Novais, A. Q. (2009). *Effect of life cycle impact assessment on the design and scheduling of a recovery network for industrial polluted waste.* Carcow, Poland: European Symposium of Computer Aided Process Engineering.

Elkington, J. (2004). *Enter the triple bottom line* (pp. 1–16). The triple bottom line: Does it all add up.

Fandel, G., & Stammen, M. (2004). A general model for extended strategic supply chain management with emphasis on product life cycles including development and recycling. *International Journal of Production Economics, 89*, 293–308.

Figge, F., & Hahn, T. (2004). Sustainable value added–measuring corporate contributions to sustainability beyond ecoefficiency. *Ecological Economics, 48*(2), 173–187.

Fleischmann, M., Krikke, H. R., Dekker, R., & Flapper, S. D. P. (2000). A characterisation of logistics networks for product recovery. *Omega, 28*, 653–666.

Ilgin, M. A., & Gupta, S. M. (2010). Environmentally conscious manufacturing and product recovery (ECMPRO): A review of the state of the art. *Journal of Environmental Management, 91*, 563–591.

Linton, J. D., Klassen, R., & Jayaramann, V. (2007). Sustainable supply chains: An introduction. *Journal of Operations Management (JOM), 25*, 1075–1082.

Kleindorfer, P. R., Singhal, K., & Van Wassenhove, L. N. (2005). Sustainable operations management. *Production and Operations Management, 14*(4), 482–492.

Krikke, H., Bloemhof-Ruwaard, J., & Van Wassenhove, L. N. (2003). Concurrent product and closed-loop supply chain design with an application to refrigerators. *International Journal of Production Research, 41*, 3689–3719.
Gomes, I., Barbosa-Póvoa, A., & Novais, A. (2011). Modeling a recovery network for WEEE: A case study in Portugal. *Waste Management, 31*, 1645–1660.
Goedkoop, M. J., Heijungs, R., Huijbregts, M., De Schryver, A., Struijs, J., & Van Zelm, R. (2009). *ReCiPe 2008, A life cycle impact assessment method which comprises harmonised category indicators at the midpoint and the endpoint level* (1st ed.). Report I: Characterisation.
Guide, D. R., & Van Wassenhowe, L. N. (2002). The reverse supply chain. *Harvard Business Review, 80*(2), 25–26.
Guillen-Gosalvez, G., & Grossmann, I. E. (2010). A global optimization strategy for the environmentally conscious design of chemical supply chains under uncertainty in the damage assessment model. *Computers & Chemical Engineering, 34*, 42–58.
Grossmann, I. E. (2004). Changes in the new millennium: product discovery and design, enterprise and supply chain optimization, global life cycle assessment. *Computers & Chemical Engineering, 29*, 29–39.
Hagelaae, G., & van der Vorst, J. G. A. (2002). Environmental supply chain management: using life cycle assessment to structure suppy chains. *International Food and Agribusiness, 4*, 399–412.
Hassini, E., Surti, C., & Searcy, C. (2012). A literature review and a case study of sustainable supply chains with a focus on metrics. *International Journal of Production Economics, 140*, 69–82.
Hugo, A., & Pistikopoulos, E. N. (2005). Environmentally conscious long-range planning and design of supply chain networks. *Journal of Cleaner Production, 13*, 1471–1491.
Jayaraman, V., Guide, V. D. R., & Srivastava, R. (1999). A closed-loop logistics model for remanufacturing. *Journal of the Operational Research Society, 50*, 497–508.
Meckenstock, J., Barbosa-Póvoa, A. P. F. D., & Carvalho, A. C. (2015). The wicked character of sustainable supply chain management: Evidence from sustainability reports. *Business Strategy and the Environment*. doi:10.1002/bse.1872.
Neto, J. Q. F., J.M. Bloemhof-Ruwaard, J. A. E. E., van Nunen J. A., & Heck, E. (2008). Designing and evaluating sustainable logistics networks. *International Journal of Production Economics, 111*, 195–208.
Neto, J. Q. F., J.M. Bloemhof-Ruwaard, J. A. E. E., van Nunen J. A., & Heck, E. & Spengler, T. (2009). A methodology for assesssing eco-efficiency in logistics networks. *European Journal of Operational Research, 193*, 670–682.
Matos, S., & Hall, J. (2007). Integrating sustainable development in the supply chain: the case of life cycle assessment in oil and gas and agricultural biothechnology. *Journal of Operations Management, 25*(6), 1083–1102.
Mota, B., Gomes, I., Carvalho, A., & Barbosa-Póvoa, A. P. (2015). Towards supply chain sustainability: Economic, environmental and social design and planning. *Journal of Cleaner Production, 15*, 14–27. doi:10.1016/j.jclepro.2014.07.052.
Parrish. B. (2009), Sustainability-driven entrepreneurship: Principles of organization design. *Journal of Business Venturing*.
Pinto-Varela, T., Barbosa-Póvoa, A. P. F. D., & Novais, A. Q. (2011). Bi-objective optimization approach to the design and planning of a supply chain: Economic versus environmental performances. *Computers & Chemical Engineering, 35*(8), 1454–1468.
PRÉ Consultants b. v., *Eco-indicator 99 methodology report*, PRÉ Consultants b.v. Amersfoort, Netherlands, 2001.
Realff, M. J., Ammons, J. C., & Newton, D. (2000). Strategic design of reverse production systems. *Computers & Chemical Engineering, 24*, 991–996.
Rubio, S., Chamorro, A., & Miranda, F. (2008). Characteristics of the research on reverse logistics. *International Journal of Production Research, 46*(4), 1099–1120.
Salema, I., Barbosa-Póvoa, A. P. F. D., & Novais, A. Q. (2006). A warehouse based design model for reverse logistics. *Journal of Operational Research Society, 57*(6), 615–629.

Salema, I., Barbosa-Póvoa, A. P. F. D., & Novais, A. Q. (2007). An optimization model for the design for a capacitated multi-product reverse logistics network with uncertainty. *European Journal of Operations Research, 179*, 1063–1077.
Salema, I. G., Barbosa-Póvoa, A. P. F. D., & Novais, A. Q. (2009). A strategic and tactical model for closed-loop supply chains. *OR Spectrum, 31*, 573–599.
Salema, M., Barbosa-Povoa, A., & Novais, A. (2010). Simultaneous design and planning of supply chains with reverse flows: A generic modeling formulation. *European Journal of Operational Research, 203*, 336–349.
Seuring, S., & Müller, M. (2008). From a literature review to a conceptual framework for sustainable supply chain management. *Journal of Cleaner Production, 16*, 1699–1710.
Seuring, S. (2013). A review of modeling approaches for sustainable supply chain management. *Decision Support Systems, 54*, 1513–1520.
Soylu, A., Oruc, C., Turkay, M., Fujita, K., & Asakura, T. (2006). Synergy analysis of collaborative supply chain management in energy systems using multi-period MILP. *European Journal of Operational Research, 174*, 387–403.
Stindt, D., & Sahamie, R. (2014). Review of research on closed loop supply chain management in the process industry. *Flexible Services and Manufacturing Journal, 26*, 268–293.
Turkay, M., Oruc, C., Fujita, K., & Asakura, T. (2004). Multi-company collaborative supply chain management with economical and environmental considerations. *Computers & Chemical Engineering, 28*, 985–992.
UNGC—United Nations Global Compact. (2013) Global corporate sustainability report 2013. Accessed March 2014.
You, F., Tao, L., Graziano, D. J., & Snyder, S. W. (2012). Optimal design of sustainable cellulosic biofuel supply chains: Multiobjective optimization coupled with life cycle assessment and input–output analysis. *AIChE Journal, 58*, 1157–1180.
Zeballos, L., Gomes, M., Barbosa-Povoa, A. P., & Novais, A. Q. (2012). Addressing the uncertain quality and quantity of returns in closed-loop supply chains. *Computers & Chemical Engineering, 47*, 237–247.
Zeballos, L., Méndez, C. A., Barbosa-Povoa, A. P., Novais, A. Q. (2014) Multi-period design and planning of closed-loop supply chains with uncertain supply and demand. *Computers And Chemical Engineering, 66*, 151–164.
Zhou, Z., Cheng, S., & Hua, B. (2000). Supply chain optimization of continuous process industries with sustainability considerations. *Computers & Chemical Engineering, 24*, 1151–1158.
WCED, World Commission on Environmental and Development. (1987). *Our common future*. New York: Oxford University Press.

Enterprise Systems and Supply Chains

Johan Magnusson and Andreas Nilsson

Abstract Enterprise systems in the form of Enterprise Resource Planning (ERP), Customer Relationship Management (CRM) and Supply Chain Management (SCM) systems have become the very backbone of organizational value creation. Through this, they both act as enablers and constrainers for processes such as supply chain management. This chapter focuses on establishing the basics to the current development of technology and business.

Keywords Enterprise systems · Supply chain management

1 Introduction

Enterprise systems in the form of standardized, enterprise-wide information systems have proliferated the market for the past decades. At the core of this technology lies the digitalization of the full scope business processes, from sales to manufacturing and procurement. Since these systems have now become an integral part of the infrastructure for value creation in most firms, all processes have become amalgamated with the information technology support.

From this basic premise follows the necessity for students of business processes to have an understanding of the underlying technology. Students of e.g. supply chain management need to be aware of the intimate interplay between the very processes that they are involved in and the technology. In addition to this, much of the work with optimizing supply chains is dependent upon both the analytical

J. Magnusson (✉) · A. Nilsson
Department of Applied IT, Chalmers and University of Gothenburg, Gothenburg, Sweden
e-mail: johan.magnusson@gu.se

A. Nilsson
e-mail: andreas.nilsson@ait.gu.se

J. Magnusson
Westerdal Oslo School of Art, Communication and Technology, Oslo, Norway

A.P. Barbosa-Póvoa et al. (eds.), *Optimization and Decision Support Systems for Supply Chains*, Lecture Notes in Logistics,
DOI 10.1007/978-3-319-42421-7_2

capabilities of corporate software (i.e. proprietary, standardized commercial-off-the-shelf software) and the available data (i.e. data residing in enterprise systems as well as external sources of both a structured and non-structured nature).

This chapter on enterprise systems and supply chain management explores the links between the processes and technology, as well as opens up for a critique of the presently dominating perspective on technology as merely supporting. In contrast, technology is regarded as both supporting and at the same time regulating work.

To aid the student, we have enclosed an excerpt from the book "Enterprise System Platforms: Transforming the Agenda" (Magnusson and Nilsson 2014). The book is published by Studentlitteratur in Lund, Sweden and is intended for students and practitioners with an interest in better understanding what this interplay between technology and business is, was and is currently evolving towards.

2 The Six Forces of IT

In Fig. 1, we summarize the underlying forces that are descriptive for these past few years' evolution. As seen in the figure, we have not addressed the intricate interplay and influence between the identified forces, but instead focused on describing them in a sequential manner. Albeit an interesting aspect of the development, we have refrained from this level of analysis for the sake of readability and to avoid the logical fallacy *post hoc ergo propter hoc*.

Fig. 1 The six forces of IT

2.1 The Digitalization of Everything

> This new technology, let's call it Information Technology.
>
> Leavitt and Whisler (1958, p. 41).

Since the 1950s, the type of technology referred to as *information technology* (IT) has been introduced into more and more aspects of social life. Stemming from machinery intended for calculating large amounts of data (large in the relative use of the term) for the military, government and business, the technology was early on identified as having fundamental implications for various strands of life.

In relation to business, one of the prominent thinkers in relation to IT at the time, Harvard Professor of Accounting John Dearden (1922–1989) offered his vision of what the end-state of the current level of technological development was.

> … the more information available, the better the decision. This end is to be accomplished by having vast amounts of data stored in a computer memory, by having this information constantly updated by point-of-action recorders, by having direct interrogation of the data stored in the computer's memory available to the executive, and by having immediate visual display of the answer.
>
> Dearden (1964, p. 128.)

Readers of today may find it hard to understand the extent to which this line of thinking was radical, but needless to say we see substantial evidence of Dearden's foresight in today's business environment. The rise of business intelligence (BI) solutions and the developments surrounding Big Data seem almost eerily hyphenated in the quote. In addition to this, Dearden also highlights one of the underlying drives behind this strive for total and real-time information, i.e. rational decision making. If every physical action and event is recorded, we will be in a position where all of our decisions, in theory, will be informed and hence freed from irrational guess works. When action becomes digitalized information, we can handle it in a rational manner.

As noted by March and Olsen's (1986) dominating garbage can model and Simon's notion of *bounded rationality*, the very concept of rationality warrants further attention. Human decision-making is, perhaps by definition and default, more a-rational or quasi-rational than rational? It is not the intent of this book to take a stance in relation to this question. Interestingly though, we would advise the reader to carefully take stock of the *intent* of the information technology currently available. We believe that herein lies a proverbial conflict between the design and use of information technology. This is of particular interest when we consider recent developments such as the rise of solutions for 'Prescriptive analytics', with the systems themselves actually making the decisions.

Digitalization does not, however, stop at the updating of information in a 'computer's memory' as noted by Dearden. It has vaster consequences, and according to some, it also brings with it the blurring of boundaries between the physical and the non-physical, between work as we have known it for years and work as we (perhaps) will know it in the future. Researchers such as McAfee and Brynjolfsson (2008) note that the main attribute of information technology is the digitalization of the very atom of business itself. This atom is the process; or in other words the workflow that in aggregated form constitutes business.

With the rise of enterprise systems such as Enterprise Resource Planning (ERP) systems during the 1990s, processes are hard-coded into the very fabric of information technology *en masse*. They are manifested through blueprints, and the only way to execute a process is through the information technology interface. The upside to this is the rapid deployment of process related innovations, such as e.g. an optimal way to handle returns for a global consumer goods firm. Through information technology, this could, once again in theory, be implemented overnight leading to a homogenization and optimization of the entire firm's global process for return handling.

Through the works of researchers such as Clayton Christensen, we have started to understand the disruptive implications of IT for social and corporate life. Through phenomena such as the Internet of Things (or the Internet of Everything), more and more of what we have seen as separated from IT is rapidly becoming entangled in technology. Industries where IT traditionally has been seen as an administrative or production technology are undergoing shifts where either the entire product or service is digitalized (such as in the music industry), or IT is becoming a substantial part of the product (such as in the automobile industry) or service (such as in the management consulting industry). This shift brings with it new entrants and competitors, echoing the premonition put forth by Michael Porter in the 1980s (1985).

2.2 The Standardization of the Unique

Before we address the issue of standardization, we need to clarify what we are actually referring to when we refer to Standardization. Perhaps the best way to do this is to clarify what standardization *is not,* to eliminate some of the common misconceptions related to this term. Standardization *does not* mean that all things are the same. In this manner, the existence of standardized processes within a firm *does not* mean that all processes are homogenous. It *does not* mean that all configurations are equal, or that variants of processes cannot be found. We refer to standardization along the lines of the Capability Maturity Model (CMM), where processes are standardized if they are described following a previously agreed upon notation and nomenclature. Hence, following technology standards does not mean that we may only use the predefined applications of technology, but that we must stick to certain rules and regulations in our application of technology. Related back to CMM, the final level of maturity in terms of standard compliance simply means that we use the same language in describing our objects.

IT has traditionally been geared towards economies of scale, where the organization agrees upon the 'best' way of configuring a process and then selects or develops an information system to support this in an economically rational manner. This has given rise to organizations striving for what they often refer to as *global processes*, or global process templates, ensuring that the organization as a whole follows the same process for e.g. financial reporting. The underlying rationale behind this is that following a global template: we ensure both economies of scale and internal communication, agreeing to one set of definitions and a common workflow. This has given rise to the birth of Shared Service Centers, or centralized 'factories' handling the entire organization's administrative needs related to specific processes. The core of this idea is that there should not be any individualized customizations to the process, but that everybody needs to agree on what is set in the global template.

This poses an interesting question with regard to the tradeoff between economies of scale versus economies of scope. In economies of scale, the striving for efficiency is highlighted, whereas in economies of scope the striving for effectiveness and adaptability is emphasized. Global processes could be regarded as a concrete example of how organizations strive for economies of scale, at the potential cost of economies of scope. This brings forth the issue of agility, and the increased demands on firms not to consider competitive advantage as something than can be sustainable over time. This is in sharp contrast with previous conceptions of strategic management, in which 'sustainable competitive advantage', where the resources were not easily imitated, was seen as the optimal state for an organization. Today, we are more and more turning our attention to issues such as dynamic capabilities, agility and continuous change, with 'sustainable' competitive advantage being a contradiction in terms.

Enterprise systems come with a predefined set of processes, geared for creating global processes, similar to what Upton and Staats (2008) refer to as the building of a cathedral. The key to the cathedral is the issue of knowing exactly what you want, and the inability to use the structure before it is completed. Once completed, it will stand for hundreds of years, supporting the identified requirements of the past. The tradeoff between efficiency and effectiveness (or scale and scope) is clear: how can organizations with a constant need for reconfiguring their processes and business models achieve both efficiency and effectiveness through standardized processes?

2.3 The Commoditization of Processes

One of the general trends is the shift of things traded towards commodities. In this process of commoditization, products or services that were previously customized are repackaged into commodities.

Being packaged as commodities brings with it the promise of reducing the cost involved in making a transaction on the market in question (often referred to as the *transaction cost*) through decreasing the time that a customer has to spend in selecting what she is intent on buying. At the same time, it reduces the cost involved in switching between vendors (often referred to as the *switching cost*). Since the commodity is packaged in a similar fashion by different vendors, the buyer can, at least theoretically, exit her relationship with a current vendor and engage with a new one, this without the characteristics of the commodity changing noticeably.

While the commoditization of products has been going on for quite some time, the commoditization of services has only recently been addressed on a larger scale. In 2005, Professor Thomas Davenport published a paper in the Harvard Business Review on how *processes* were currently undergoing commoditization (Davenport 2005). As Davenport argued, the rise of process standards such as COBIT, SCOR and ISO14001 bring with them a common nomenclature and language to describe the processes. This can be regarded as a first step in the commoditization of processes.

When firms can describe their processes in a manner that can be understood by actors on the outsourcing market, processes will become a commodity traded like any other one on an open market. If we as a firm, for instance, were to describe our supply chain process following the SCOR methodology, we could more easily communicate our process specifications, the expected level of performance and the cost to external parties. If the said process were to be sourced to a lesser price from an actor on the open market than from internal resources, then we could choose to outsource this particular process to the vendor most meeting our requirements in terms of price and quality.

This phenomenon of sourcing processes (or sub-processes) from external vendors is referred to as Business Process Outsourcing (BPO) and has throughout the past couple of years seen a radical increase in market size. At the same time as this development can be understood from an economic point of view, it raises several questions as to the very nature of the firm. What actually constitutes the firm as we understand it? What constitutes the boundaries of the firm?

The impacts of the commoditization of processes have also been highlighted for firms working outside of traditional industry. Christensen et al. (2013) advocate an upcoming commoditization of the services offered by management consultants, where we see firms such as McKinsey and Associates packaging elements of their previous delivery for faster and more efficient delivery. The potential of technologies such as crowdsourcing of analysis (e.g. Kaggle.com), prescriptive analytics (e.g. Ayata) and self-service BI (e.g. Tableau Software) shifts a large portion of what was previously supplied by the management consultants as a complete package. In other words, new technology-induced solutions are disrupting the very firms that have recommended firms to invest in the said technology.

2.4 The Consumerization of Technology

Any user having been exposed to corporate IT while at the same time using consumer IT will testify to the sharp contrasts between the two. Consumer IT has experienced a drastic growth during the past couple of decades, creating a chasm within a technology that initially was designed for the corporate realm. IT was initially so complex and costly that any consumer-directed application of it would be commercially impossible. It was a technology designed for professionals, be they accountants, physicians or officers.

During the 1980s, a new wave of technology started to proliferate the market. The IT industry was starting to re-frame itself towards end users, through innovations such as desktop computing and the spinoffs this technology brought with it. In the following decade, the Internet was introduced as a medium through which communications could be made even less costly and available for a larger part of the community. In the early years, vendors strived to create value in what initially best could be referred to as an empty room. Connecting people (which coincidentally was the byline of Nokia, one of the dominant cellphone vendors at the time) and achieving network externalities, along the lines of what is commonly referred to as Metcalfe's law: the utility of a network increases exponentially with every added node (Gilder 1993).

As the incumbent vendors saw the massive potential of a market for consumer IT, they were at the same time distraught about how to feed the rapid onslaught of innovations back into the market for corporate IT. Having established themselves on a market where they currently had a strong position, they were adamant towards making radical changes to their existing solutions. Hence, the direct effects of a massive increase in innovation for consumer IT did not spill over to the corporate side.

At the same time, new challenging vendors of corporate IT saw massive potential in the new technology being introduced for consumers. In it they saw the necessary prerequisites for transforming the graphical user interface (GUI) and the way we as users consume technology. With the introduction of software as a service (SaaS) as a delivery model for software, challenging vendors often offer the intended customers the option of trying their solution out for free (Fig. 2).

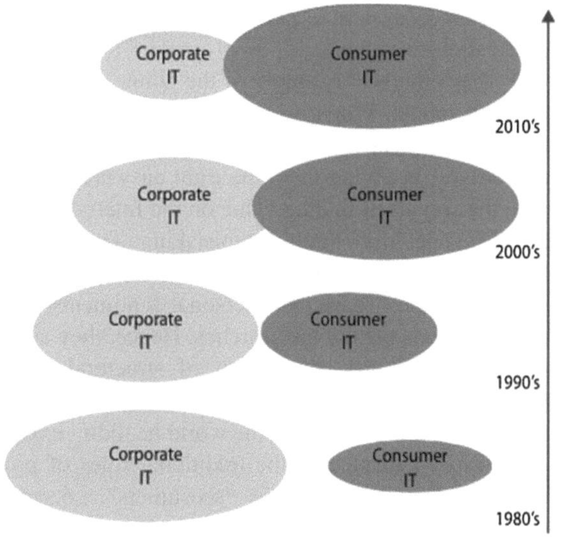

Fig. 2 Illustration of how consumer and corporate IT have changed over the past few decades

This challenges the previously so dominant position of the incumbent vendors, and opens up for a blending of consumer and corporate IT through the introduction of a new line of products and services. At the same time it creates a market for vendors that sell their solutions directly to the end user, not necessarily attending to often centralized models of IT procurement for the customer. This in turn creates a situation where a larger and larger part of an organization's total IT spending is becoming decentralized, with a lack of corporate cost control as a direct consequence. This phenomenon is referred to as "Shadow IT", and according to prominent industry analysts, the proportion of IT spending that falls within the category of Shadow IT will reach as much as 90 % by 2018. We will address this in more detail in the Sect. 2.6.

2.5 *The Co-creation of Value*

In the shift towards cloud-based delivery models such as SaaS, we have a hard time seeing the value of our infrastructure in the books. Instead of procuring a resource, we subscribe to it, and hence we have no depreciation and no book value of the service in question. A service (such as SaaS) is in this respect something that generates value momentarily on usage, and not something that we can place in our inventory if we wish to use it later on instead of now. In addition to this, the value of a service is created in the meeting of the client and provider/service, and hence not simply created but rather *co-created.*

In research, there has long been a strong tradition of looking into strategic alliances and other inter-organizational collaborations. Despite this, there seems to be a lack of understanding in terms of which mechanisms exist between organizations involved in co-creation. This is noted by Sarker et al. (2012), in their study of how an ecosystem of partners surrounding a large ERP vendor is involved in the co-creation of value.

Perhaps the best example of the allure of co-creation can be found in Google's business model. When we ask our students (in a highly unscientific manner) about who Google's customers actually are, the answer is predominantly the students themselves, in giving them the right answers to their searches and supplying them with the service of finding value on the Internet. In response to this answer, we then ask if perhaps they might be mistaking "customer" for "factory worker", with the following logic: the students generate revenue by advertising for Google. At the same time, they share their personal sentiments and information while optimizing the algorithms behind the searches. Hence, they are involved in both direct revenue generation and the accruement of structural capital for Google. As noted by Baudelaire (2011) in his famous poem Spleen in Paris: "The greatest trick the devil ever pulled was convincing the world he didn't exist." Co-creation brings with it the possibility of displacing the traditional roles of producer and consumer, creating what is often referred to as "prosumers", i.e. actors involved with the parallel production and consumption of a product or service.

An alternative to this is the growth of what is commonly referred to as the *sharing* or *collaborative* economy. According to The Economist (2013), this grew to a $26 billion market during 2013. Examples such as Airbnb, a service that supports sub-letting private lodgings to travellers, or Uber, a ride-sharing service, thrive on platform logic in connecting private owners of lodging and transportation with consumers wanting to share resources. These firms are competing with traditional channels such as hotels and taxis, as well as toppling the previous composition of services offered by e.g. travel agencies through co-creation.

2.6 The Disintegration of Systems

> Turning and turning in the widening gyre
>
> The falcon cannot hear the falconer;
>
> Things fall apart; the centre cannot hold;
>
> Mere anarchy is loosed upon the world.
>
> W.B. Yeats, The Second Coming

In the early 1950s, novelist and Nobel laureate Chinua Achebe described the social transformation taking place in a small, Nigerian rural village. With the inflow of new norms and influences, the social fabric of everyday life started to change, resulting in both wonderful new possibilities and the loss of that which once was.

The situation described in the novel is one of disintegration. Defined as the "shift from larger to smaller pieces", disintegration is also central to understanding what has been happening for the past 30-odd years within the IT industry, particularly in relation to enterprise systems.

Software development has undergone several shifts since its birth in the 1950s (some would argue that the development of software goes back even further, to Thomas von Neumann or even Karl Leibnitz). Central to the change in development methodology and software architecture has been the striving for re-use and loose coupling of code snippets. These snippets were first referred to as functions, later on objects and currently services. The services of today stem from a new approach to building systems first introduced in the late 1990s and referred to as Service Oriented Architecture (SOA).

SOA stipulates that a system should have the ability to use and issue services from and to (from the system itself) external systems. Hence, a system designed following a SOA approach should both be able to share its own code and calculations with other systems, and not be self-sufficient in terms of code. Hence, the exchange of services between systems (often referred to as *web services*) becomes a signifying mark for SOA based systems.

In stipulating and agreeing on the standards for the exchange of the said services and the possibility of utilizing internet protocols instead of proprietary network solutions as the medium for exchanging services, the cost of integration has radically decreased on a per unit basis. The implications of SOA on IT in general and enterprise systems particular have been an increased disintegration, somewhat paradoxically due to the increase of inter-system integrations. Since systems no longer need to be self-sufficient in terms of code and functionality, the previously so strong *raison d'être* for monolith solutions has become obsolete. With integration being standardized, it is now possible to combine services from a number of systems into one single set of functionality, without a disproportionate increase in cost.

Hence, the disintegration of systems is expressed in the increase of cross-dependencies between systems and the possibility of satisfying functional requirements through combining existing services from a multitude of systems. The consequences of this can be seen in such diverse phenomena as SaaS and Shadow IT, and a mean decrease in project scope for software development. Much in line with what Chinua Achebe described in "Things fall apart", disintegration brings with it both substantial possibilities, but also significant problems for users and vendors alike.

Further Reading

Baudelaire, C. (2011). *Le spleen de Paris: petits poèmes en prose*. Le livre de poche.

Chinua, A. (1958). *Things fall apart*. London: United Press.

Christensen, C. M., Baumann, H., Ruggles, R., & Sadtler, T. M. (2006). Disruptive innovation for social change. *Harvard Business Review, 84*(12), 94.

Christensen, C. M., Wang, D., & van Bever, D. (2013). Consulting on the Cusp of Disruption. *Harvard Business Review, 91*(10), 106–114.

Cohen, M. D., March, J. G., & Olsen, J. P. (1972). A Garbage can model Organizational Choice. *Administrative Science Quarterly, 17*(1), 1–25.

Davenport, T. H. (2005). The coming commoditization of processes. *Harvard Business Review, 83* (6), 100–108.

Dearden, J. (1964). Can management information be automated? *Harvard Business Review, 42*(2), 128–135.

Gilder, G. (1993). Metcalfe's law and legacy. *Forbes ASAP, 13*.

Jensen, M. C., & Meckling, W. H. (1976). Theory of the firm: Managerial behavior, agency costs and ownership structure. *Journal of Financial Economics, 3*(4), 305–360.

Leavitt, H. J., & Whisler, T. L. (1958). Management in the 1980's. *Harvard Business Review, 36* (6), 41–48.

March, J. G., & Olsen, J. P. (1986). Garbage can models of decision making in organizations. *Ambiguity and command*, 11–35.

McAfee, A., & Brynjolfsson, E. (2008). Investing in the IT that makes a competitive difference. *Harvard Business Review, 86*(7/8), 98.

Porter, M. E. (1985). *Competitive strategy: Creating and sustaining superior performance*. New York: The free.

Porter, M. E., & Millar, V. E. (1985). How information gives you competitive advantage. *Harvard Business Review*, 1–15 (July–August).

Sarker, S., Sarker, S., Sahaym, A., & Bjørn-Andersen, N. (2012). Exploring value co-creation in relationships between an ERP vendor and its partners: a revelatory case study. *MIS Quarterly, 36*(1), 317–338.

Simon, H. A. (1979). Rational decision making in business organizations. *The American Economic Review, 69*(4), 493–513.

The Economist (2013). http://www.economist.com/news/leaders/21573104-internet-everything-hire-rise-sharing-economy. Retrieved April 12, 2016.
Tiwana, A., & Konsynski, B. (2010). Complementarities between organizational IT architecture and governance structure. *Information Systems Research, 21*(2), 288–304.
Upton, D. M., & Staats, B. R. (2008). Radically simple IT. *Harvard Business Review, 86*(3), 118.
Woodard, C. J., Ramasubbu, N., Tschang, F. T., & Sambamurthy, V. (2013). Design capital and design moves: The logic of digital business strategy. *MIS Quarterly, 37*(2), 537–564.
Zittrain, J. L. (2006). The generative internet. *Harvard Law Review, 119*, 1974.

Downstream Oil Products Distribution Planning: A Portuguese Case Study

Nuno Mota, Susana Relvas and Jorge Gonçalves

Abstract In the actual worldwide environment, the oil industry faces fierce and growing competition. In this context, oil supply chains should be studied in order to improve their efficiency, while remaining flexible to successfully handle certain types of contingencies, such as lack of products or fleet unavailability. At the distribution level, tankers are commonly used in the downstream activities to transport the derivative products from distribution centres to service stations, comprising the secondary distribution level. This operation is usually short-term scaled to meet final consumers' demand. However, the availability and proper sizing of a fleet to perform the required distribution can be a complex problem due to the fact that demand is known on short notice and the distribution network may include hundreds of demand points to be satisfied. In this chapter, the T2S.opt—Tank to Station Optimizer—decision support tool is presented. T2S.opt addresses the fleet distribution planning problem under normal and abnormal operational scenarios. The optimal planning covers short-term solutions and minimizes operational costs. A Mixed-Integer Linear Programming (MILP) was developed and implemented to be used through a proper user interface, giving origin to T2S.opt. The software was used to schedule the secondary distribution of oil products of GalpEnergia in Portugal.

Keywords Oil supply chain · Downstream · Secondary distribution · Planning · Mixed-integer linear programming

N. Mota · S. Relvas (✉)
Centro de Estudos de Gestão CEG-IST, Instituto Superior Técnico, Universidade de Lisboa, Av. Rovisco Pais, 1049-001 Lisbon, Portugal
e-mail: susana.relvas@tecnico.ulisboa.pt

J. Gonçalves
PETROGAL SA, Rua Tomás da Fonseca, Torre C, Lisbon, Portugal

A.P. Barbosa-Póvoa et al. (eds.), *Optimization and Decision Support Systems for Supply Chains*, Lecture Notes in Logistics,
DOI 10.1007/978-3-319-42421-7_3

1 Introduction

In the actual worldwide environment, the oil industry faces fierce and growing competition. In this context, oil supply chains should be analysed in order to improve their efficiency, while remaining flexible to successfully handle certain types of contingencies, such as lack of products or fleet. At the downstream secondary distribution level, tankers are commonly used, due to their flexibility to transport the derivative products from distribution centres to service stations (MirHassani 2008). This operation is usually short-term scaled so as to meet final consumers' demand. However, the availability and proper sizing of a fleet to perform the required distribution can be a complex problem due to the fact that demand is known on short notice and the distribution network may include hundreds of demand points to be satisfied.

The optimal fleet allocation solution would require a Vehicle Routing Problem (VRP), or one of its variants such as the Capacitated VRP, to be solved in short periods of time over a geographically large area with several demand locations (service stations country wide or region wide). This problem is known to be NP hard and even small instances are difficult to solve. Since this is a daily exercise for many distribution companies, other approaches are required. On the other hand, real world tools used to this end usually are based on geographic knowledge and experience of schedulers to perform such allocation. The gap between an optimal distribution planning tool and what is currently used in companies leaves the opportunity for the academic community to focus on feasible and good solutions in short scheduling periods. Moreover, decision support tools using such type of solutions are of great support when dealing with different types of operational contingencies.

In this work we present a decision support system (DSS), named Tank-to-Station Optimizer (T2S.Opt), which addresses the fleet distribution planning problem under normal and abnormal operational scenarios. The system in study considers a network of oil products distribution centres that supply service stations on a daily basis. Distribution centres are supplied from refineries. The DSS is built so as to consider the distribution operation over a short-term time horizon (one to few days) minimizing operational costs. This DSS uses a MILP model which is solved with a free solver (GLPK). The DSS includes a user interface and a flexible architecture that can be fully customized, and uses the MILP-based solution strategy reported in Mota (2012). The proposed methodology is tested in a real world case study of a Portuguese company—GalpEnergia—which distributes oil products nationwide.

In the following section a literature review in the field of oil supply chains is presented. Section 3 describes the proposed mathematical model and a brief overview of T2S.opt. Section 4 presents the case study as well as the results obtained by using T2S.opt. Finally, Sect. 5 encloses the conclusions and future work.

2 Literature Review

The oil supply chain is commonly classified under three main segments: upstream, midstream and downstream. The first one encompasses the crude oil exploration and transportation up to refineries, the midstream involves the refining operations and lastly the downstream segment is concerned with the physical distribution of oil products to an extensive and diverse retail sector or to the petrochemical industry (An et al. 2011).

The transportation operations that take place throughout this supply chain may use diverse transportation types, as vessels, train, truck or pipelines. Vessels, sometimes even reaching VLCC—Very Large Crude Carriers—sizes, are used in crude transportation. Inland supply may also require either pipelines or train, depending on the quantity, distance and frequency of supply. On the downstream side, the distribution is usually broke down to primary and secondary (Fig. 1), where distribution centres (or depots) play a central role in managing oil products. Mainly in the secondary distribution level, trucks play a central role due to their enhanced performance in milk run type distribution. The remaining transportation modes are more frequent in the primary distribution level.

Pipelines have been extensively used in the oil supply chain over the last 40 years. Despite the initial investment, their operational cost is reduced when compared to other transportation types (Relvas et al. 2006; Herrán et al. 2010). In alternative, vessels are the most indicated for harbour connected nodes and large quantities (MirHassani 2008).

Planning in a supply chain might be different according to several aspects as performance measures and decision variables change and dependent on the business. According to An et al. (2011), five planning levels may be distinguished: strategic, tactical, operational, integration of tactical and operational, integration of strategic and tactical. Gayialis and Tatsiopoulos (2004), among other authors, chooses to distinguish only the first three levels of planning mentioned above.

Strategic planning uses aggregated information and regards long term decisions as investments and dimensioning of the logistics network. This may include locating and determining the capacity of the depots, physical flows, sourcing strategy, among other decisions (Gayialis and Tatsiopoulos 2004). The objective

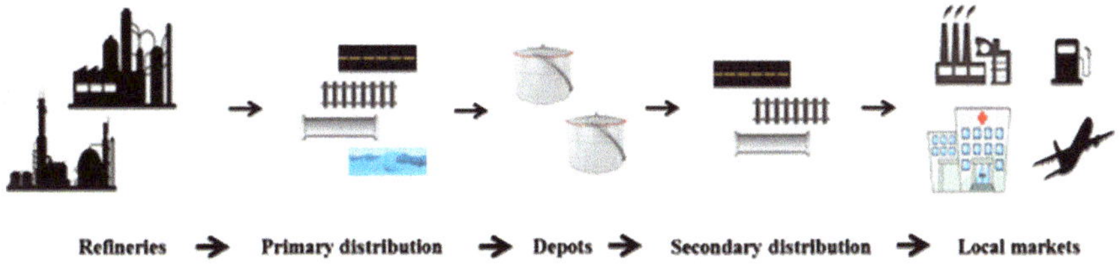

Fig. 1 Downstream oil supply chain

functions are normally related to costs and flexibility of the network that is being modified or created (see Mota 2012).

Tactical planning regards medium-term decisions as inventory levels, production planning and distribution related to an existing network. The objective functions normally take into consideration the costs (or profit) and the responsiveness to the customer.

Operational planning considers detailed information and short-term decisions, such as daily repetitive operations, and concentrates in optimizing specific points of the network (Gayialis and Tatsiopoulos 2004). It may include transportation scheduling, detailed resource allocation, production decisions or distribution and routing problems as the vehicle routing problems and its variants (see Mirabi et al. 2010; Nussbaum et al. 1997). Typically, these types of problems are NP hard to solve and heuristics are proposed to solve them (see Mirabi et al. 2010).

The integration of the strategic and tactical models typically handles tactical decision variables, such as production and/or distribution scheduling, levels of inventory, and strategic decisions, like physical allocation, sizing infrastructures and/or transports, among others (Mota 2012). These models focus more in the downstream operations (An et al. 2011).

The integration of operational and tactical models, when compared with operational models, tend to include more than one operation and are more concentrated in the production activity (see Timpe and Kallrath 2000; Al-Othman et al. 2008). Given the focus of the present work, the integration of these two planning levels will allow to answer to our main research questions. However, most of the literature relies in normal operation scenarios and few methods are available to address contingency planning in distribution scenarios.

There are several approaches that can be used when dealing with uncertainty, risk and subsequent reactive planning. Uncertainty is related to the non-deterministic character of the variables under scrutiny.

Although it is generally consensual that planning taking into account uncertainty or adopting risk management strategies might bring considerable benefits, there are few approaches applied to the oil supply chain. Adhitya et al. (2007) propose a framework based on a rescheduling heuristic strategy to manage failures in a refinery. However, the oil supply chain may face numerous uncertainty factors or abnormal scenarios. To this end, decisions at different levels may be implied when finding solutions for these situations.

Real world downstream distribution scenarios have to cover several customer locations sourced from different refineries or distribution centres. Furthermore, under non-regular scenarios (e.g. lack of stock of a given product) tactical decisions related with allocation of customers to sourcing locations may arise at the operational level. Therefore, our aim is to develop a problem representation where the tactical level is integrated with the operational level of decision, where two major decisions are to be defined: allocation of customer locations to sourcing nodes and number of supply vehicles to be shipped from sourcing nodes.

3 Problem Definition

The problem in study consists of a distribution network with multiple depots and refineries. Each refinery supplies several depots with the respective products. The number, location and product availability of the depots are known. A fleet of trucks with different capacities and limited availability per truck type is used to transport the products from depots to customers (e.g. municipalities). In this way, either single customers or aggregated customers may be dealt within the proposed formulation. The time required to travel between network points (from depots to customers) is also deterministic and known. Each truck cannot visit multiple regions in one trip. This is mainly due to safety constraints in the routing of vehicles loaded with oil derivatives when they have to visit more than one customer location. Since stability issues may arise, frequently only one customer is visited or a restricted number of customer that are located within a short distance. Thus, the problem at hand differs from capacitated VRPs since the focus is customer allocation and demand fulfilment and fleet capacity planning.

Given this setting, we aim to determine the optimal distribution planning of refined products from depots to customers in short-term time horizons that minimizes distribution costs, which include transportation costs. Penalties for unmet demand are also accounted for.

3.1 Mathematical Formulation

The proposed solution is divided in two phases. In the first phase we aim designing the delivery routes from a depot to customers at a minimum cost. In case it is not possible to meet all the demand, due to fleet or/and product constraints, a second phase returns the minimum time that it would be required to supply that demand. Basically, the second phase problem has no fleet limitations.

The proposed formulation uses the following notation consisting of indexes, sets, parameters and variables.

Indexes:

i Depots;
j Customers;
k Products;
p Tankers;
t Time periods;

Sets:

$i \in I$ Set of all depots;
$j \in J$ Set of all customers;
$k \in K$ Set of products;
$p \in P$ Set of tankers;

$t \in T$ Set of time periods;
$R_{k,p}$ Set that relates products with tanker to transport it;
$S_{k,p,i}$ Set that relates products with tankers and depots;

Parameters:
$A_{i,k,t}$ Initial inventory of product k, at depot i, at time t (m^3);
$C_{i,j}$ Unitary distribution cost between depot i and customer j (€);
$Cam_{i,t,p}$ Available shifts from tanker p, in depot i, at time period t;
$Dem_{j,k,t}$ Demand of customer j of product k at time period t (m^3);
$IF^{min}_{i,k,t}$ Minimum inventory of product k, in depot i, at time period t (m^3);
$IF^{max}_{i,k,t}$ Maximum inventory of product k, in depot i, at time period t (m^3);
$Min_{i,j}$ Travelling time between depot i and customer j (h);
M Big-M number related with the order of magnitude of variable $Y_{i,j,k,t}$;
Pen Penalty per unit of volume not transported between depot i and customer j (€), defined as $Pen > MaxC_{i,j}$;
$TP_{i,t}$ Maximum flow from depot i at time period t (m^3);
$TR_{i,k,t}$ Refinery flow to depot i, from product k, at time period t (m^3);
V_p Transportation capacity of tanker p (m^3);

Non-negative Variables:
$X_{i,j,k,t}$ Flow of product k transported from depot i to customer j at time period t;
$Xn_{j,k,t}$ Demand of product k from customer j not satisfied at time period t;
$IF_{i,k,t}$ Final inventory of product k, in depot i at time period t;
$Y_{i,j,k,t}$ Demand of product k from customer j satisfied in the second phase and allocated to depot i at time period t;
$Yn_{j,k,t}$ Demand of product k from customer j not satisfied in the second phase at time period t;

Integer Variables:
$n_{i,j,t,p}$ Number of vehicles from depot i to customer j, from tanker p, at time period t;

Mathematical Model—1st Phase:

$$\text{Min } Z = \sum_t \sum_k \sum_j \left(\sum_i (C_{i,j} \times X_{i,j,k,t}) + Pen \times Xn_{j,k,t} \right) \quad (1)$$

$$Xn_{j,k,t} = Dem_{j,k,t} - \sum_i X_{i,j,k,t}, \quad \forall j, k, t \quad (2)$$

$$\sum_j \sum_k X_{i,j,k,t} \le TP_{i,t}, \quad \forall i, t \quad (3)$$

$$A_{i,k,t} + TR_{i,k,t} - \sum_j X_{i,j,k,t} = IF_{i,k,t} \quad t = 1, \forall i, k \tag{4}$$

$$IF_{i,k,t-1} + TR_{i,k,t} - \sum_j X_{i,j,k,t} = IF_{i,k,t} \quad t > 1, \forall i, k \tag{5}$$

$$\frac{\left(\sum_k X_{i,j,k,t}\right)}{V_p} = n_{i,j,t,p} \forall (k,p) \in R_{k,p}, \quad \forall j, t, I \tag{6}$$

$$\sum_i \sum_j \left(n_{i,j,t,p} \times Min_{i,j}\right) \le Cam_{i,t,p} \forall (k,p,i) \in S_{k,p,i}, \ \forall t \tag{7}$$

$$IF^{min}_{i,k,t} \le IF_{i,k,t} \le IF^{max}_{i,k,t}, \quad \forall i, k, t \tag{8}$$

$$IF_{i,k,t}, X_{i,j,k,t}, Xn_{j,k,t} \ge 0, \ \ n_{i,j,t,p} \in \mathbb{N}, \quad \forall i \in I, \forall j \in J, \forall k \in K, \forall t \in T \tag{9}$$

The objective function (1) minimizes the transportation costs between depots and customers (first term), as well as the penalties in case of unsatisfied demand (second term).

Equation (2) acts as a soft constraint that determines unmet demand. Equation (3) limits throughput of each depot in each time period. Equation (4) defines the final inventory at the beginning of the time horizon and Eq. (5) the final inventory for the subsequent time periods. Equation (6) ensures that the cargo transported by one vehicle type (and taking into account the product to be transported) is a multiple of the capacity of that type—due to safety conditions for the transportation. Equation (7) constraints the fleet transportation capacity, given different trucks (or tanks) and depots. Constraints in Eq. (8) limit the product inventory per depot and time period. Finally, Eq. (9) is a non-negativity constraint for flows and inventory.

Mathematical Model—2nd Phase:

$$\text{Min } Z = \sum_t \sum_i \sum_j \sum_p \left(Min_{i,j} \times n_{i,j,t,p}\right) + \sum_t \sum_k \sum_j \left(M \times Yn_{j,k,t}\right) \tag{10}$$

The second phase returns as a solution what are the fleet requirements to meet unfulfilled demand of the first phase. In this case, the objective function minimizes the required time to fulfil the new demand (first term), by allocating the unmet demand of the 1st phase to new depots, as well as it minimizes the penalties when such allocation is not possible (second term).

$$Yn_{j,k,t} = \sum_i Y_{i,j,k,t} - Xn_{j,k,t} \quad \forall j, k, t \tag{11}$$

$$\mathrm{IF}_{i,k,t} \geq \sum_j \mathrm{Y}_{i,j,k,t} \quad \forall i, k, t \tag{12}$$

$$\sum_j \sum_j \mathrm{Y}_{i,j,k,t} \leq \mathrm{TP}_{i,t} - \sum_j \sum_k \mathrm{X}_{i,j,k,t} \quad \forall i, t \tag{13}$$

$$\frac{\left(\sum_k \mathrm{Y}_{i,j,k,t}\right)}{\mathrm{V}_p} = n_{i,j,t,p} \quad \forall (k,p) \in \mathrm{R}_{k,p}, \forall j, t, I \tag{14}$$

$$\mathrm{Yn}_{j,k,t}, \mathrm{Y}_{i,j,k,t} \geq 0, \quad n_{i,j,t,p} \in \mathbb{N}, \quad \forall i \in I, \forall j \in J, \forall k \in K, \forall t \in T \tag{15}$$

Equation (11) allocates the unmet demand $\mathrm{Xn}_{j,k,t}$, calculated in the 1st phase, to the nearest possible depot. Constraint in Eq. (12) ensures that there is sufficient inventory in the depots allocated in Eqs. (11) and (13) limits the throughput of each depot at time t. Equation (14) restricts the flow of products to the exact capacity of each tank. Finally, Eq. (15) is the domain definition constraint.

4 Case Study and Results

4.1 Case Study Summary

The object of this study consists of the downstream fuel distribution planning of four refined petroleum products: gasoline, diesel, heating oil (HO) and Jet Fuel (JF). This activity is performed by Petrogal, a subsidiary company of the Group GalpEnergia, from this point onwards referred as Galp.

The distribution activity includes six depots, as one can see in the Table 1 with different capacities and products being stored, which satisfy a total of roughly 270 Portuguese municipalities. Figure 2 illustrates depots' locations, as well as crude oil flow. The numbers presented in Fig. 1 illustrate the actual distribution regions covered by outsourced carriers.

The data used to test the proposed approach is referent to 2011.

Table 1 Average product availability by depot (m^3)

Depots	Gasoline	Diesel	HO	JF
Boa Nova	10,000	10,000	–	700
Real	–	400	10,000	–
CLC	10,000	10,000	–	10,000
Mitrena	800	800	10,000	–
Sines	10,000	10,000	–	1225
Aveiro	–	–	–	–

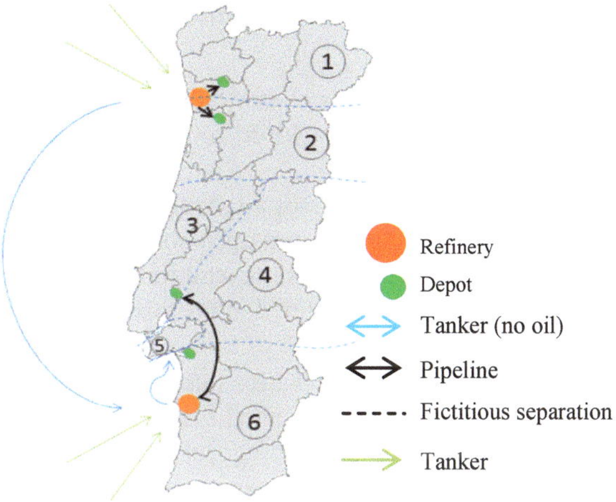

Fig. 2 Galp's simplified national supply chain

Within the six depots, Aveiro is not currently in use. In the case of Jet Fuel, Boa Nova stores typically around 700 m^3, Sines roughly 1225 m^3 and CLC has practically no fuel limitation.

The daily national consumption of gasoline, diesel and HO (aggregated consumption) in 2011 was analysed through a histogram map and the authors concluded that the most frequent consumption quantities are situated in the ranges of 0–6000 m^3/day and 8000–9000 m^3/day. Days with higher consumptions may reach volumes of 10,000–13,000 m^3/day, but with a frequency below to 20 %. The fuel transportation is outsourced to five carriers. These carriers are allocated to depots, and each carrier manages the amount of trucks to perform the transportation service. In the present scenario, Table 2 illustrates how many trucks are available per day in each depot (1 shift equals 8 h of transportation and some trucks perform 2 shifts per day).

The maximum capacities of each normal truck tank are 32 and 35 m^3 in the case of JF. Typically, the latter type of truck performs an average of 5 deliveries per day and the former an average of 3 deliveries.

The contingencies to which the distribution activity is exposed can be broadly categorized between absences of product or absence transport capacity. In the case of Jet Fuel there is a clear limitation of the carrying capacity given the scarcity of this type of truck.

Table 2 Daily trucks and shifts available by depot

	Trucks	Shifts	JF trucks	JF shifts
Boa N. + Real	51	78	(pipeline)	–
CLC Aveiras	40	49	9	18
Mitrena	6	9	–	–
Sines	20	34	(railway)	–
Aveiro	–	–	–	–

4.2 Case Study Results and T2S.Opt Decision Support Tool

The proposed mathematical models were implemented in the T2S.opt application. This application was developed to interpret model solutions in an effective and interactive way.

This application possesses a specific architecture, which integrates an Excel and Access file with the proposed open source optimization Solver GLPK. The Excel file is the user interface where several functionalities are available. The usage of this application will be exemplified in the results section, since all figures are generated by T2S.opt.

Several contingency scenarios were implemented in the T2S.opt application. The objective of the results presented is to determine the impact of closing any of the existing depots and the subsequent solution of reallocating the available fleet (where the depot was closed). First, we are going to present a normal distribution scenario in order to compare it with the contingencies referred. The closure of distribution centres will be illustrated with the examples of CLC and Sines.

4.2.1 Normal Distribution Scenario

This scenario contemplates an ordinary day of distribution. To fulfil 7048 m^3 of diesel, 2326 m^3 of gasoline and 64 m^3 of HO, the distribution operation consumes a total of 77,815 min, 47,076 km and a distribution cost of 118,890 €. In terms of JF, for a daily supply of 65 deliveries (2275 m^3) from CLC to the Lisbon airport, it requires 8515 min of transportation time and an expense of 32,258 € in transportation costs. The areas of influence of the five depots considered are shown in Fig. 3.

The illustrated areas are consistent with the current distribution scheduling. Thus, in contingency, these areas are expected to move in the direction of the closed depot, to minimize the unfulfilled demand.

4.2.2 Abnormal Scenario: Closure of CLC

This scenario includes the closure of CLC distribution centre. 71 % of the total demand was satisfied, which means that 2784 m^3 were unfulfilled. The regions whose demand was not satisfied are illustrated in Fig. 4a. In this scenario were consumed 57,221 distribution minutes and the total distribution cost was of 76,908 €. The unitary distribution cost was 11.55 €/m^3 transported, a 7.94 % decrease when compared with the normal distribution scenario.

However, one can minimize the negative impacts by reallocating the available fleet in CLC to the correct depots. Figure 4b, in the following figure, illustrates the impact of reallocating the non-used fleet. With the solution computed by the second phase of the mathematical model one allocated 5 shifts to Aveiro and 44 to Sines. This measure made the unsatisfied demand decreased 69 %, when compared to the

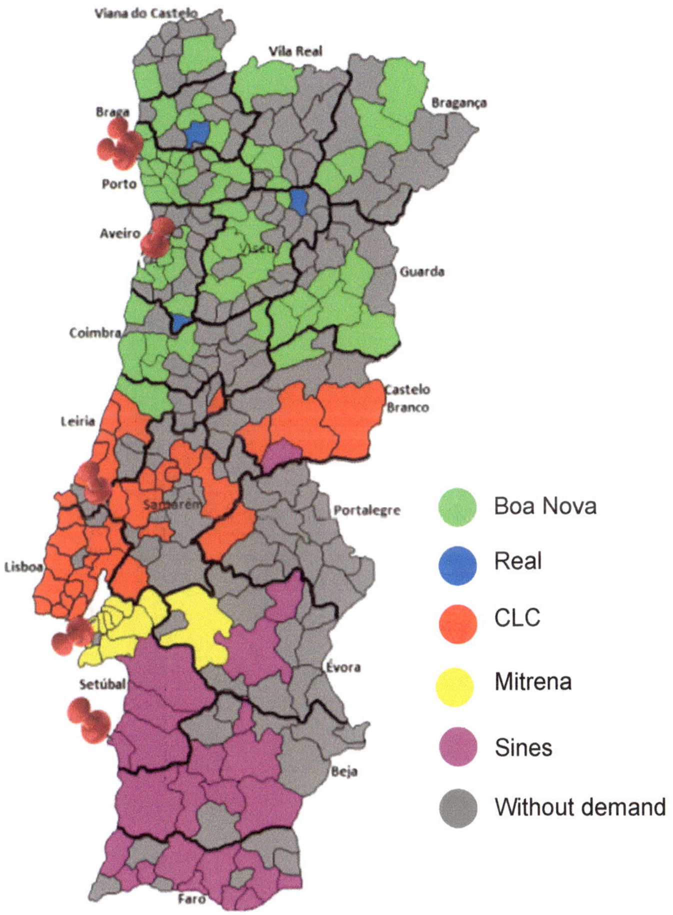

Fig. 3 Normal distribution scenario depot influence areas

scenario where no fleet was reallocated, to a total of 864 m^3. In this scenario the unitary cost of distribution increased 0.31 %, from 12.59 to 12.63 €/m^3.

4.2.3 Abnormal Scenario: Closure of Sines

This scenario includes the closure of Sines. 86 % of the total demand was satisfied, which means that 1312 m^3 were unfulfilled. Figure 5a shows the unsatisfied municipalities. In this case, 65,147 min were used for a total distribution cost of 107,108 €. This gives a unitary cost of 13.18 €/m^3 transported, a 4.67 % increase when compared with the normal distribution scenario.

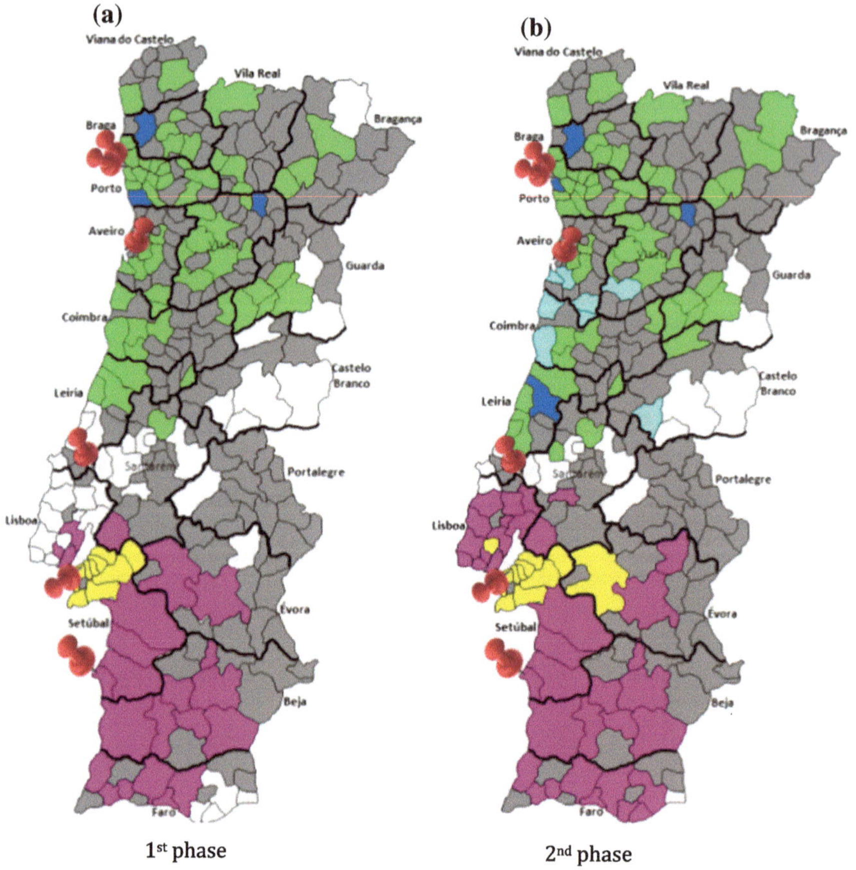

Fig. 4 Abnormal scenario: closure of CLC

In the reallocation scenario one considered 7 shifts in the Mitrena and 27 in the CLC. With this measure one decreased the unsatisfied demand to 384 m^3, a 71 % drop when compared with the no reallocation scenario. The unitary cost of distribution increased 10.64 %, from 12.59 to 16.01 €/m^3 (when compared with the normal scenario).

4.2.4 Abnormal Scenario: Fleet Unavailability

Correct contingency impact identification will not only allow us to clarify the most critical contingencies, as well as aligning the expectations and goals of the current programming team in the company in study. Typically, the most ordinary contingency is related to the lack of fleet. So we decided to study the impact of an unexpected 25 and 50 % fleet unavailability. The demand used in this study was the same of the previous sections.

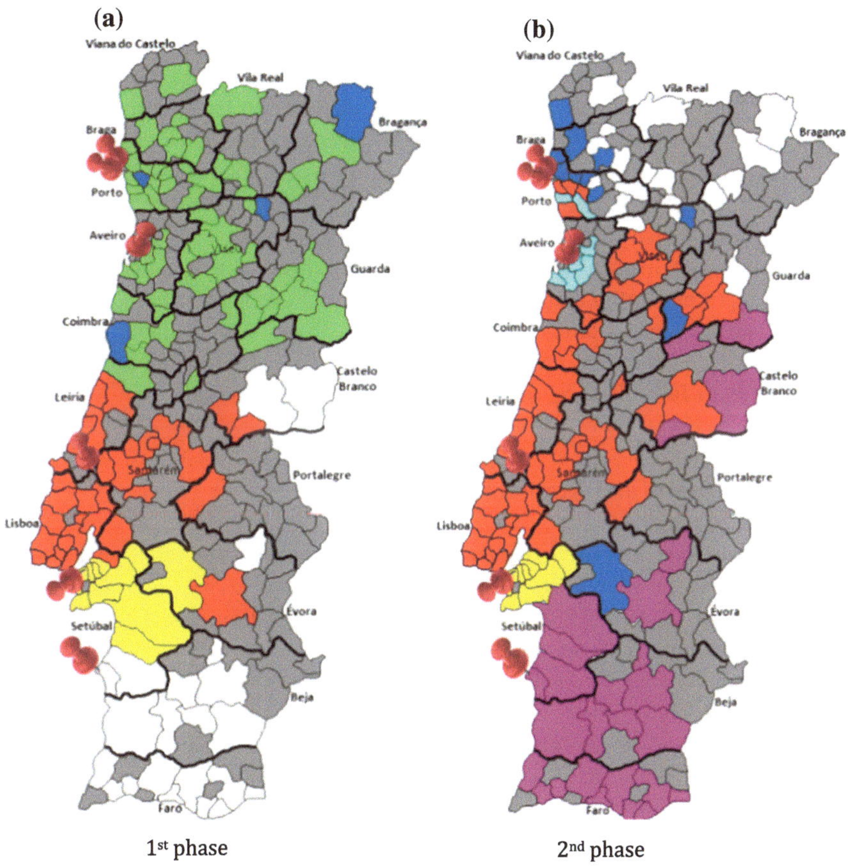

Fig. 5 Abnormal scenario: closure of sines

Table 3 Abnormal scenario fleet unavailability: impact of 25 and 50 % fleet unavailability

Scenarios	Unmet demand (m^3)	Shifts needed	U.C. of distribution (€/m^3)	U.C. of transport (€/m^3)
Normal	0	0	12.59	8.90
50 % Boa N.	960	35	11.82	8.01
50 % CLC	832	26	11.92	8.80
25 % Boa N.	320	14	12.26	8.57
50 % Sines	284	11	12.21	8.42
25 % CLC	256	11	12.27	8.89
50 % Mitre.	96	4	12.60	9.11
25 % Sines	64	3	12.50	8.80
25 % Mitre.	0	0	12.66	9.15

Table 3 summarizes each scenario impact. The first column introduces the scenario and the fleet's percentage drop in the respective depot. Subsequently, it is presented the amount of unmet demand per scenario, the shifts needed to fulfil the unmet demand and the unitary costs (U.C.) of distribution and transport.

As one can see, and in line with contingency scenarios previously studied, a drop of 50 % of the fleet in Boa Nova and CLC depots has serious consequences on the fulfilled demand. Note also that, for example Boa Nova, in order to supply the 960 m^3 of unfulfilled demand, it would require 35 additional shifts. That is, each delivery takes an average of 1.2 shifts, or approximately 9 h, to be performed. Another important fact is that the unitary cost of transportation, and distribution, is higher than the normal unitary costs only when there is unavailable fleet in the Mitrena depot. This means that longer distances have to be covered with origin in other depots to fulfil demand otherwise supplied by Mitrena depot.

4.2.5 Fleet Resize and Reallocation

In this section we propose an alternative to the current fleet configuration, by redistributing the available fleet. The goal is to increase savings and mitigate the current operational contingency potential of Galp's operation. For this purpose we are going to use the T2S.opt application.

In order to evaluate the best configuration that suits both goals we constructed a sample of 18 daily demands based in 2011s demand. For each, we calculated the optimum fleet allocation to depots, in terms of shifts, using the first phase of the mathematical model. Table 5 summarizes the results obtained, where the column named average (Avg.) represents the fleet configuration proposed as an average of the values obtained within the sample.

Comparing the results obtained with the current fleet configuration (5th column of Table 4) there is a positive 10 shift deviation (6th column). However, it was calculated a deficit of 4 shifts in Mitrena. In terms of distribution costs, the

Table 4 Fleet resize and reallocation results

Depot	Min.	Max.	Avg.	Now	Dev.
Boa Nova + Real	65	81	75	78	+3
CLC	36	45	41	49	+8
Mitrena	12	15	13	9	−4
Sines	25	38	31	34	+3
Total shifts	138	179	160	170	+10

Table 5 Daily cost comparison of fleet configurations

Costs	Actual config. (€)	Proposed config. (€)
Transport	82,854	83,335
Product	33,509	32,645
Total	116,363	115,980

Table 6 Computational results

Scenarios		Computational time (CPU s)		Relative gap (%)	
		1st Ph.	2nd Ph.	1st Ph.	2nd Ph.
Normal	Average	758.9	0.1	0.0	0.0
	Maximum	2901.5	0.1	0.0	0.0
	Minimum	1.3	0.1	0.0	0.0
Contingency	Average	12,953.9	0.1	1.0	0.0
	Maximum	47,945.5	0.1	5.1	0.0
	Minimum	184.4	0.0	0.0	0.0

proposed configuration represents an average daily gain of 383 € in relation to the current configuration.

The decomposition of the average distribution costs for the current fleet and for the proposed fleet scenario is presented in Table 5.

As one can see, both alternatives to the current fleet configuration represent an improvement in the average distribution costs.

4.3 Computational Results

Both models were implemented using the programing language of GUSEK and solved in an Intel® Core(TM)2 Duo with 2.67 GHz and 4 GB RAM computer, using GLPK solver. The stopping criteria are either the optimal solution determination or reaching the memory limit.

Table 6 resumes computational data in two categories, normal distribution scenarios and contingency scenarios, in average terms, for the sake of simplicity.

One can conclude that the computational effort varies significantly with the scenario being tested in the first model phase. However, contingency scenarios, which are more restrictive, tend to consume more time and maintain higher relative gaps.

The first phase of the model is characterized by 21,360 variables, 6405 are integer variables, and 8631 equations. The second phase is characterized by 21,360 variables, 6405 are integer variables, and 8605 equations. This model size is applicable to all scenarios, which consisted in a single day of time horizon (one time period).

5 Conclusions and Further Work

This paper proposes a decision support tool, based in an exact model that aims at minimizing operational costs of the secondary distribution operation. The solution procedure includes a MILP model solved using a free solver—GLPK—and managed through a user interface, the T2S.opt application.

The proposed MILP model allows the integration of tactical and operational decisions, such as the determination of areas of influence per distribution centre (frequently a tactical decision) and the daily distribution planning (operational decision). This model flexibility allows obtaining a solution to support the distribution operation decisions even when abnormal scenarios occur (as stated, lack of product or lack of fleet) as well as for fleet sizing (tactical decision).

A current limitation of the proposed mathematical models lies in the fact that a vehicle is only able to perform one delivery per trip. The introduction of a condition, in future developments, that allows freight to make more than one delivery per truck will enable more accurate and real results.

Acknowledgments The authors thank GalpEnergia all the support in terms of problem definition and data gathering. The authors thank Fundação para a Ciência e Tecnologia (FCT) for financial support, under the project PEst-OE/EGE/UI0097/2011.

References

Adhitya, A., Srinivasan, R., & Karimi, I. A. (2007). Heuristic Rescheduling of crude oil operations to manage abnormal supply chain events. *AIChE Journal, 53*(2), 397–422.

Al-Othman, W. B. E., Labadidi, H. M. S., Alatiqui, I. M., & Al-Shayji, K. (2008). Supply chain optimization of petroleum organization under uncertainty in market demands and prices. *European Journal of Operational Research, 189*(3), 822–840.

An, H., Wilhelm, W. E., & Searcy, S. W. (2011). Biofuel and petroleum-based fuel supply chain research: A literature review. *Biomass and Bioenergy, 35*(9), 3763–3774.

Gayialis, S. P., & Tatsiopoulos, I. P. (2004). Design of an IT-driven decision support system for vehicle routing and scheduling. *European Journal of Operational Research, 152*, 382–398.

Herrán, A., de la Cruz, J. M., & de Andrés, B. (2010). A mathematical model for planning transportation of multiple petroleum products in a multi-pipeline system. *Computers & Chemical Engineering, 34*, 401–413.

Mirabi, M., Fatemi, G. S., & Jolai, F. (2010). Efficient stochastic hybrid heuristics for the multi-depot vehicle routing problem. *Robotics and Computer Integrated Manufacturing, 26*(6), 564–569.

MirHassani, S. A. (2008). An operational planning model for petroleum products logistics under uncertainty. *Applied Mathematics and Computation, 196*, 744–751.

Mota, N. (2012). Análise de cenários para definição de planos de contingência na distribuição secundária de produtos petrolíferos, Master Thesis in Industrial Engineering and Management, Instituto Superior Técnico, Universidade Técnica de Lisboa, Portugal (in Portuguese).

Nussbaum, M., Sepulveda, M., Cobian, A., Gaete, J., Parra, E., & Cruz, J. (1997). A fuel distribution knowledge-based decision support system. *Omega, 25*(2), 225–234.

Relvas, S., Matos, H. A., Barbosa-Póvoa, A. P. F. D., Fialho, J., & Pinheiro, A. S. (2006). Pipeline scheduling and inventory management of a multiproduct distribution oil system. *Industrial and Engineering Chemistry Research, 45*, 7841–7855.

Timpe, C. H., & Kallrath, J. (2000). Optimal planning in large multi-site production networks. *European Journal of Operational Research, 126*(2), 422–435.

Reverse Logistics: Concept, Evolution and Marketing Challenges

Sergio Rubio and Beatriz Jiménez-Parra

Abstract Reverse logistics is a research area focused on the management of the recovery of products once they are no longer desired or can no longer be used by consumers, in order to obtain an economic return through reuse, remanufacturing or recycling. In recent years, reverse logistics has become a matter of strategic importance in the context of the Supply Chain Management. In this chapter, an introduction to the concept and evolution of reverse logistics is provided; from a very basic approach related to the recycling channels to the current concept of closed-loop supply chain. Then, we describe some of the challenges for reverse logistics in the next years, especially those related with the marketing of the recovered products, namely competition issues in remanufacturing, the cannibalization problem, the purchase intention of consumers of remanufactured products, and their perceptions and willingness to pay for this sort of products. Research in reverse logistics from a demand/market perspective is nascent and there exist challenging opportunities to keep developing this topic.

Keywords Reverse logistics · Remanufacturing · Marketing · Returned products

A previous and shortened version of this manuscript was published in the International Journal of Engineering Business and Management, an Open Access Journal (Rubio and Jiménez-Parra 2014).

S. Rubio (✉)
Department of Business Management & Sociology, Universidad de Extremadura, Badajoz, Spain
e-mail: srubio@unex.es

B. Jiménez-Parra
Department of Management & Business Economics, Universidad de León, León, Spain

A.P. Barbosa-Póvoa et al. (eds.), *Optimization and Decision Support Systems for Supply Chains*, Lecture Notes in Logistics,
DOI 10.1007/978-3-319-42421-7_4

1 Introduction

Reverse logistics has attracted the attention of companies and academia in the last two decades. Some examples in the business world can be found in firms such as Apple, Canon, Caterpillar, Dell, Electrolux, Hewlett-Packard, or IBM which include reverse logistics practices in their operational processes. This growing interest is also evidenced by the increase in the level of related activities in leading sectors such as the transport sector, the consumer electronics sector, the textile sector, and the press and media, to name but a few (Verstrepen et al. 2007). In a similar vein, from an academic point of view, reverse logistics has been a topic of growing interest in recent years, as can be observed in the increasing trend of the number of published articles (Agrawal et al. 2015a), monographies and books. An example of this growing interest can be found in several literature reviews published very recently (Souza 2013; Agrawal et al. 2015a; Govindan et al. 2015).

Reverse logistics is a research area focused on the management of the recovery of products once they are no longer desired or can no longer be used by consumers, in order to obtain an economic return through reuse, remanufacturing or recycling (Flapper et al. 2005). Some of the most relevant issues analysed have focused on aspects related to the analysis of the flow of goods from the consumer back to the producer or to the recovery entity (Thierry et al. 1995; Guide and Van Wassenhove 2001); the production planning and inventory management (Toktay et al. 2000; Yang et al. 2005); and coordination issues among different processes in the Supply Chain (SC) (Savaskan et al. 2004; Ferguson et al. 2006). Regarding gaps and new opportunities for research, issues such as the integration of different levels of decision-making and paying attention to multi objective problems (Govindan et al. 2015); development of models for forecasting product returns and risk assessment of reverse logistics implementation (Agrawal et al. 2015a); and empirical research on consumer behaviour and the market for remanufactured products (Souza 2013), deserve to be mentioned.

In this chapter, an introduction to the reverse logistics concept and its implications and challenges for the SC is presented. To this aim, an overview of the evolution of this concept is provided as well as a brief review of the main contributions in this field. In Sect. 2, we review the implications that the reverse logistics concept can generate for the SC, by analysing the main decisions made when implementing a reverse logistics system. In Sect. 3, we describe some of the challenges for reverse logistics in the next years, especially those related with the marketing issues of the recovered products. As usual, we finally provide some conclusions and insights.

2 Reverse Logistics: An Overview

According to Thierry et al. (1995) and De Brito and Dekker (2004), the main reasons or drivers for implementing a reverse logistics system are the following:

- Economic reasons: They can be classified as direct and indirect reasons. Direct reasons are related to a more efficient use of raw materials, reduction of disposal costs and creation of added value from returned products. Regarding the indirect reasons, an image of environmentally responsible behaviour and improved customer relations could be two of them.
- Legal reasons: Because of environmental legislation, many companies are held responsible for the products they produce or distribute throughout their life cycles (Extended Producer Responsibility). This primarily involves that enterprises have to develop ways for recovering the End-of-Use (EoU) products in order to recover their value (reuse, remanufacturing, recycling) or to a proper disposal.
- Social reasons: Social awareness regarding the environment had led to increasing demands (from customers, stakeholders, non-governmental organizations, etc.) for environmentally responsible behaviour by companies. Most of these "social requirements" have focused on reduction of carbon emissions and waste generation.

The concept of reverse logistics has evolved over the years (De Brito and Dekker 2004), passing through varying stages until becoming consolidated. From a very basic approach related to the raw material recycling (Ginter and Starling 1978), to the big impulse given by contributions from the field of engineering and operational research in the nineties (see for example, Fleischmann et al. 1997), until these days in which it is a fully recognised subfield of supply chain management (Guide and Van Wassenhove 2009).

During all this period, several definitions have been suggested for the concept of reverse logistics, for example:

> …the term often used to refer to the role of logistics in recycling, waste disposal, and management of hazardous materials; a broader perspective includes all relating to logistics activities carried out in source reduction, recycling, substitution, reuse of materials and disposal (Stock 1992).

> The process of planning, implementing and controlling the efficient, cost-effective flow of raw materials, in-process inventory, finished goods, and related information from the point of consumption to the point of origin for the purpose of recapturing value or proper disposal (Rogers and Tibben-Lembke 1999).

> …a process by which a manufacturing entity systematically retrieves previously shipped products or parts from the point-of-consumption for possible recycling, remanufacturing, or disposal (Dowlatshahi 2005).

Although the conception of reverse logistics dates from 1970s (see, for example, Guiltinan and Nwokoye 1974; Ginter and Starling 1978), the denomination of this term is difficult to trace with precision (Table 1). During the seventies and eighties, the scientific literature refers to terms like reverse channels or reverse flows consistently associated with recycling operations, so the definition was inspired by the movement of flows against the traditional flows in the SC; at the beginning of the nineties, the first known definition of reverse logistics is published by the Council of Logistics Management; at the end of the 1990s, reverse logistics is characterised by

Table 1 Evolution of the reverse logistics concept

Period of time	Description	Main contributions
1970s	Use of different terms ("reverse channels", "reverse flows") in conjunction with the recycling concept	• Guiltinan and Nwokoye (1974) • Ginter and Starling (1978)
1980s	Emphasis on several terms as "backwards" and "forwards" in order to highlight the existence of flows with movements in the opposite direction of the traditional flows	• Lambert and Stock (1981) • Murphy (1986) • Murphy and Poist (1989)
Early 1990s	First formal definition of reverse logistics (Council of Logistics Management)	• Stock (1992) • Pohlen and Farris (1992) • Kopicky et al. (1993)
In the late 1990s	Recognition of the important role played by "management" in the field of reverse logistics	• REVLOG (1998) • Rogers and Tibben-Lembke (1999)
From the year 2000 onwards	Holistic perspective: joint vision of forward and reverse flows. Development of CLSC concept	• Guide et al. (2003) • Lebreton (2007) • Guide and Van Wassenhove (2009) • Ferguson and Souza (2010)

Source Compiled by authors based on Dekker et al. (2004) and Carrasco (2010)

the recovery of the value of EoU products and the processes involved; in these days, a holistic view of the supply chain is proposed by considering, simultaneously, forward and reverse flows from a business perspective. From that moment on, management of both types of flows cannot be conceived separately, this is the so-called closed-loop supply chain (CLSC) management, that can be defined (Guide and Van Wassenhove 2009) as the design, control, and operation of a system to maximize value creation over the entire life cycle of a product with dynamic recovery of value from different types and volumes of returns over time. In this sense, Guide and Van Wassenhove (2009) provide an interesting analysis about the evolution of the CLSC research and use five phases to highlight the evolutionary process of this research area: (1) the golden age of remanufacturing as a technical problem, (2) from remanufacturing to valuing the reverse-logistics process, (3) co-ordinating the reverse supply chain, (4) closing the loop, and (5) prices and markets.

However, the reverse logistics definition proposed by the European Working Group on Reverse Logistics, REVLOG, will be considered throughout this chapter. This research group defined reverse logistics as "*the process of planning, implementing and controlling backward flows of raw materials, in process inventory, packaging and finished goods, from a manufacturing, distribution or use point, to a point of recovery or point of proper disposal*" (De Brito and Dekker 2004).

As it was previously mentioned, the growing interest on this topic can also be observed in the number of reviews published (for example, Prahinski and Kocabasoglou 2006; Rubio et al. 2008; Guide and Van Wassenhove 2009; Ilgin and Gupta 2010; Souza 2013; Agrawal et al. 2015a; Govindan et al. 2015), that can be checked to obtain a more detailed view of this field. Nevertheless, we can highlight a set of papers that have undoubtedly contributed to the development of this research area:

- Thierry et al. (1995) can be considered as a seminal work on product recovery management. It provides a description of the product recovery options distinguishing among them according to the reprocessing process: repair, refurbishing, remanufacturing, cannibalization, and recycling.
- Fleischmann et al. (1997) is, with no doubt, one of the key papers on this topic. It is a thorough review of the main operational research models for reverse logistics focusing on three crucial issues: distribution planning, inventory management, and production planning.
- Rogers and Tibben-Lembke (1999) clearly identified the key role of logistics in commercial returns and documented many practices.
- Fleischmann et al. (2000) provides a characterization of product recovery networks, and classifying them in three categories according to the recovery process: bulk-recycling networks, remanufacturing networks, and re-usable item networks.
- Savaskan et al. (2004) introduces the concept of CLSC and presents different structures of remanufacturing networks in order to analyse the interactions between the forward and the reverse channel decisions.
- Guide and Van Wassenhove (2009) provides an overview of the evolution of CLSC from a business perspective. It highlights some research needs related to (1) the development of more sophisticated operational research models to gather the business perspective of the problem, (2) the need of becoming more familiar with CLSC practice, and (3) the opportunity to build relationships with other disciplines such as marketing or accountability.
- Souza (2013) is a recent overview of the most relevant studies developed in the field of the CLSC management. This paper focuses on those articles which analyse strategic issues (i.e., remanufacturing decisions by Original Equipment Manufacturer—OEM, network design, etc.) and tactical issues (e.g., used product acquisition and disposition decisions). For each topic, a base model with underlying assumptions and results is presented and the author's perspective on needed research areas is argued.

Obviously, there exist other papers that contributed to the development of this topic, although mostly based on these ones.

3 Reverse Logistics in the Supply Chain Context

The relevance of the reverse logistics concept in the SC context is increasing, and, actually, it is difficult to find examples of SC without the more or less important presence of reverse logistics (Corominas et al. 2015).

Successful implementation of reverse logistics activities demands making decisions related to different hierarchical levels: strategic, tactical, and operational. From a strategic point of view, the consideration of return flows into the SC requires to make decisions about the number, location, and capacity of the corresponding facilities (collection points, sorting and inspection facilities, recovery centres, etc.), the design of the network (independent from the forward network vs. CLSC), and the technological processes to implement (transportation systems, inspection, classification and recovery technologies, etc.), among others. From a tactical point of view, decisions about the assignment of collected products to recovery centres, inventory management issues should be considered. Operational decisions are related to the collection and distribution routes, recovery options (recycling, remanufacturing, reuse), and waste management. Table 2 shows a summary of this sort of decisions according to the different activities developed in a reverse logistics system.

Table 2 Reverse logistics decisions

Activities	Strategic	Tactical	Operational
Collection	• Location, number and capacity of collection points • Lay-out • Process systems	• Assignment of End-of-Life (EoL) products to recovery centres • EoL inventory management • Transport means	• Collection batches • Collection routes • Load configuration
Inspection and sorting	• Location, number and capacity of inspection and sorting facilities • Training	• Inventory management for recoverable products • Assignment of tasks: disassembly, cleaning, restoring, etc.	• Recovery option decision: reuse, remanufacturing, recycling
Recovery	• Process systems: best available technologies • Markets for recovered products	• Inventory management for recovered products • Effects on Aggregate Production Plan	• Effects on Master Production Plan • Bill of materials
Disposal	• Disposal systems	• Inventory management for non-recoverable products • Transportation means	• Waste management

Source Adapted from Rubio (2003)

However, if we had to indicate a key decision in the implementation process of a reverse logistics system, it would necessarily be the reverse logistics network design (RLND). Alumur et al. (2012) highlight that reverse logistics network configuration is a complex problem that requires the determination of optimal localisations and capacities of collection centres, sorting centres, remanufacturing facilities and/or recycling plants. This way, RLND becomes a strategic issue in the context of the Supply Chain Management (SCM). Because of its relevance, a brief discussion about the design of a reverse logistics network is provided in the following lines.

Several contributions to this issue have provided a basic description of the Reverse Logistics Networks (RLN) by identifying commonalities among them and indicating critical elements in their design and implementation:

- Thierry et al. (1995) classify the RLN according to the recovery option given to the EoU product: (1) direct reuse—direct use and resell—, (2) product recovery management—repair, remanufacturing, refurbish, cannibalization and recycling—, and (3) disposal—incineration and landfill.
- Fleischmann et al. (2000) provide a classification based on the main characteristics observed in different business cases: (1) bulk-recycling networks—e.g., sand recycling, recycling of steel by-products, and carpet recycling—, (2) remanufacturing networks—e.g., copier remanufacturing, mobile phone remanufacturing and printed circuit board recovery—and, (3) re-usable item networks—e.g., reusable packages.
- Savaskan et al. (2004) analyse four different configurations according to their (de-)centralization degree in order to describe the interactions between the forward and the reverse channel decisions: centrally coordinated system, manufacturer collecting, retailer collecting and third-party collecting.
- Flapper et al. (2005) also describe diverse RLN by using business cases from different sectors (cosmetics, pharmacy, electronics and beverage, among others), paying special attention to other elements related to organisational, environmental, technical and economic aspects.

However, probably, the main consideration in RLND is the choice between an independent network for the collection of EoU products and a network integrated into the forward SC, resulting in a CLSC (Corominas et al. 2015). Anyway, both, an independent design (RLN) and a CLSC, are set around two critical activities, namely: (i) the collection of EoU and (ii) the recovery option (remanufacturing, reuse, recycling). Aras et al. (2010) provide a thorough analysis related to the design of RLNs with an approach aimed at identify the implications for the SCM according to (1) the right network structure (e.g., should a OEM use an existing retail network or a third-party firm to collect used products or collect directly from end users himself?), (2) the right collection strategy (for example, should the used products be picked-up from the end users or is it better to set up drop-off facilities for returns?, and (3) the role of financial incentives in the collection strategy (e.g., should financial incentives be given to promote the return of used products?, and how do financial incentives and the choice of the collection strategy influence the structure of the RLN?).

While network design for EoU products recovery is an active area of research in the SC literature, more research effort is needed (Akçali et al. 2009). In this vein, Alumur et al. (2012) highlight that RLN configuration is a complex problem that requires the determination of the optimal locations and capacities of the collection centres, sorting centres, remanufacturing facilities and/or recycling plants.

Nevertheless, the design of a RLN is based on three basic activities:

1. Collection of EoU products. According to Corominas et al. (2015), collection of EoU products can be considered the starting point of the system. The authors describe three different collection options depending on whether the collection is made directly by the manufacturer or remanufacturer, through a network of distributors and retailers, or through third-party logistics providers. In general, at this stage, some important decisions have to be adopted: (1) strategic decisions—location and quantity of recovery facilities, capacity, technology which have to be used and design of recovery facilities—, (2) tactical decisions—transportation planning system or EoU inventory management—, and (3) operational decisions—planning of collection routes for EoU products and configuration of collection batches.
2. Inspection and Classification. One of the main characteristics of the product recovery management is the uncertainty associated to the recovered products, in terms of quantity, quality and time (Thierry et al. 1995; Dekker et al. 2004). Most companies do not know how many products will be returned and, what the main conditions of these products will be so, they ignore if the quality of the returned products will be good enough to be properly recovered (remanufactured, recycled and reused). Furthermore, there is some uncertainty regarding the moment in which these products will be returned and, therefore, it is unknown if they will be on time to introduce them in the recovery process. In this way, these two activities (inspection and classification) will determine which the condition of the returned products is, so an analysis of location and capacities of sorting centres is required.
3. Recovery Process. It can be considered as the key element of a RLN due to, at this phase, the economic value of the returned product is recovered by using one of the following options:

   - Reuse. This is a recovery option which implies very basic activities to reconditioning the product (cleaning, minor repairs), that do not modify their structure or nature (see Carrasco-Gallego et al. 2012, for a detailed analysis). Some types of packaging and containers (pallets, returnable glass bottles, plastic containers, etc.) are the best examples of products to which this recovery option can be applied.
   - Remanufacturing. This option requires achieving additional activities, such as disassembly, inspection, repair, and assembly, in order to recover the value of the returned products and provide them similar characteristics (warranty, quality, performance, etc.) than the original products. In some

cases, when it is economically and technically feasible, improvements of several technological modules of products are performed, in order to ensure that remanufactured products include technological levels analogous than the original ones. Numerous examples of remanufactured products can be found in the electronic sector: laptops, printers, mobile phones, etc.
- Recycling. Throughout this last option, simply the economic value of the raw materials is recovered, so the returned product loses its identity. According to Ferguson and Souza (2010), this alternative could be especially recommended when materials can be separated economically by environmentally responsible techniques. Some examples of good candidates for recycling are: packaging material, glass, paper, plastic, aluminium, etc.

New opportunities for research in this stream can be considered, particularly those related to the empirical application that could be of immediate help to practitioners (Aras et al. 2010), or when combined with take-back legislation (Souza 2013).

4 Research Challenges for the Next Years

In this specific area of research, most of the problems analysed in the literature have been approached from the *supply side* of EoU products, by analysing the flow of goods from the consumer back to the producer or to the recovery agent, e.g., collection, recovery value (reuse, remanufacturing, recycling), inventory management, production planning, etc. (De Brito and Dekker 2004). However, there has been little work from the *demand side* about how the recovered products are re-introduced into the market after the recovery process. Issues such as the marketing of recovered products, their acceptance by consumers, the existence of new markets for these products and how these markets can be developed, which marketing strategies are best suited for this purpose, or what type of consumer should be targeted are some of the main concerns in the analysis of the reverse logistics from a demand point of view (Michaud and Llerena 2011; Jiménez-Parra et al. 2014; Agrawal et al. 2015b). There is a broad consensus that one of the challenges for CLSC research in the coming years is the need to examine in depth its relationships with the market and consumers (Guide and Van Wassenhove 2009; Atasu et al. 2010; Subramanian and Subramanyam 2012; Souza 2013), so in this section a brief description of the main challenges of research for reverse logistics from the *demand side* is provided.

According to Subramanian and Subramayam (2012), recent reviews of CLSC research have highlighted the need for an empirical treatment of market factors. In this sense, Guide and Van Wassenhove (2009) also claim for more interdisciplinary research with marketing and accounting areas to validate assumptions that many of the CLSC models are based on, in order to "*keep the business model perspective rather than optimizing an isolated part of the problem*". As Atasu et al. (2010)

recognise "*the marketing aspects of remanufacturing are largely unexplored by academic research*", so this can be a good moment for beginning to explore some of those aspects.

Although a more detailed review can be found in Subramanian and Subramanyam (2012) and Agrawal et al. (2015b), we here provide a classification of the main contributions focused on analysing some particular marketing issues in the context of the CLSC research, by grouping them by sub-topic, namely competition in the remanufactured product market, cannibalization, purchase intention, willingness to pay (WTP) for remanufactured products, as well as issues related to the perception of consumers of these types of products (Table 3).

4.1 Competition

Is one of the first research issues focused on the market as a key element that can have an effect on the performance of the CLSC, for which Majumder and Groenevelt (2001), Debo et al. (2005) and Atasu et al. (2008) can be considered as basic references in the study of this topic and its implications for the different participants in the remanufactured product market. In this sense, for example, Atasu et al. (2008) suggest that direct competition between OEMs may have a significant impact on the profitability of remanufacturing. In particular, they argue that remanufactured products may help the OEM compete for more price sensitive consumers, who would otherwise be interested in demanding low price original products from other OEMs. Furthermore, the authors state that "*remanufacturing is more beneficial under competition than under monopoly even in the absence of the green segment*" (price sensitive consumers). Debo et al. (2005) suggest that manufacturers that also have remanufacturing operations may benefit from managing both new and remanufactured products. They propose the idea of selling new products below unit costs in order to generate a supply of remanufactured products, on which the profit is made. Additionally, from these authors' point of view "*a decrease in the unit remanufacturing cost may lead to an increase in the new product sales volume, to supply remanufactured products in response to increased demand for them*". In the same vein, Majumder and Groenevelt (2001) study price competition between an OEM and a local remanufacturer (third-party remanufacturer, 3PR), and the effect of different strategies on the competitive prices and quantities in the market, as well as the players' profits. From a methodological point of view, this can be considered as a quantitative sub-topic which has been mainly analysed by using optimization techniques and game theory approaches. Maybe, other methodological approaches closer to the business perspective (case studies, lab experiments) could be a right complement to the classical operational research methods used in these studies. Recent research papers related to this topic are considering new elements of interest for future research in this sub-topic, such as how the product quality level and the benefits of remanufacturing depend on the party (OEM or 3PR) doing the remanufacturing (Örsdemir et al. 2014); or which

Table 3 Reverse logistics from a demand perspective

Topic	Author/s	Design-methodology-approach	Results-insights	Opportunities for further research
Competition	Majumder and Groenevelt (2001)	Game theory	Behaviour of OEMs when faced with competition from local remanufacturers	• Quality level of remanufactured products • Different cooperation modes between remanufacturing competitors • More qualitative methodologies
	Debo et al. (2005)	Dynamic optimization	Address the integrated market segmentation and production technology choice problem in a remanufacturing setting	
	Atasu et al. (2008)	Optimization	Competition, market growth, and the proportion of the green segment have a significant direct impact on the remanufacturing decision	
Cannibalization	Atasu et al. (2010)	Research paper (qualitative)	Remanufacturing does not always cannibalize original product sales. Remanufacturing can create additional value to OEMs	• Effects of cannibalization on consumer products • Effects of cannibalization on commercial products • Empirical research
	Guide and Li (2010)	Experimental lab-field auctions	Explore the potential for market cannibalization of original product sales by remanufactured versions	
Purchase intention	Wang et al. (2013)	Structural equation model	Purchase intention of consumers of remanufactured automobile spare parts	• To explore the purchase intention of consumers in a wider market segment • Studies based on representative samples of consumers
	Jiménez-Parra et al. (2014)	Structural equation model	Purchase intention of consumers of remanufactured laptops	
	Gaur et al. (2015)	Grounded theory (qualitative)	Purchase intention of consumers of remanufactured products who have been relocated to a society where remanufactured products are promoted	
Willingness to pay (WTP)	Essoussi and Linton (2010)	Survey	Consumers' WTP premium price for recycled products is product specific and perceived functional risk appears to have a statistically significant impact on consumer purchase decisions	• To explore the WTP in different products and industries, in order to a better understanding of the different results observed • Analysis of information as a determinant of the WTP • The role of branding and OEM image
	Michaud and Llerena (2011)	Experimental auctions	Consumers tend to value the remanufactured product less than the conventional. No evidence that consumers are willing to pay a premium for the remanufactured (green) product. Information as a key factor for marketing strategy	
	Ovchinnikov (2011)	Focus Group. Conjoint Analysis. Survey	Consumers would be more willing to pay for remanufactured products if they had a clear information about terminology used in this market and about product's history	

(continued)

Table 3 (continued)

Topic	Author/s	Design-methodology-approach	Results-insights	Opportunities for further research
Consumer perceptions of remanufactured products	Hazen et al. (2012)	Structural equation model	Ambiguity inherent in the remanufacturing process provokes consumers' perception of poorer quality and reduces their willingness to buy remanufactured items	• To explore the effects of information and consumer knowledge across different brand, product categories and markets • To explore the influence of OEM versus 3PR remanufactured products on the consumer perception across different products and markets
	Subramanian and Subramanyam (2012)	Regression analysis	Seller reputation and remanufacturer identity are key factors on quality perception of consumers of remanufactured products	
	Wang and Hazen (2015)	Structural equation model	Influence of consumer knowledge (quality, cost and green attribute) of remanufactured products on perceived value and perceived risk associated with purchase decisions	
	Agrawal et al. (2015b)	Behavioural experiments	The perceived value of remanufactured products and the identity of the remanufacturer can influence the perceived value of new products	

mode of cooperation between OEM and 3PR, outsourcing or authorisation, is the most appropriate (Zhou et al. 2016).

4.2 *Cannibalization*

In the context of the CLSC, "*cannibalization*" occurs when the purchase of a remanufactured product displaces the sale of an original product (Atasu et al. 2010). According to Guide and Li (2010), many sales managers are reluctant to introduce remanufactured products because they fear that original product sales can be cannibalized by remanufactured products. Actually, cannibalization has been considered as one of the main concerns of marketing and sales managers to implement remanufacturing activities. However, there is little real knowledge about the effects of cannibalization on the marketing activities and many times all the managers' concerns are based on personal experience or conventional wisdom, rather than on empirical evidence (Atasu et al. 2010; Guide and Li 2010). In this sense, Atasu et al. (2010) point out that remanufacturing does not always cannibalize original product sales and if so, the additional profits of remanufacturing can overweight the cannibalization costs. For that reason, these authors remark that several important aspects, such as composition of the market (types of consumer segments: functionality-oriented, newness-conscious and green consumers), the proper use of the price strategies, the competition, and the supply of remanufacturable products over the product life cycle, must be perfectly understood. For this same issue, Guide and Li (2010) use a novel procedure in order to determine the consumers' WTP for remanufactured products and, at the same time, to assess the effect of cannibalization of new product sales. Specifically, they performed an analysis of online auctions for a particular category of products (power tools) intended for both Business-to-Consumer (B2C) market and Business-to-Business (B2B) market, and they found that the risk of cannibalization was minimal in the B2C market, while in the second case (B2B market), there was an evidence of potential cannibalization. These novel approaches based on lab experiments are welcomed, because more empirical research is needed (Souza 2013). This sub-topic of research is usually focused on the analysis of one particular product, so further research should continue to explore the effects of cannibalization for other products in both consumer and commercial settings (Guide and Li 2010).

4.3 *The Purchase Intention for Remanufactured Products*

A third set of contributions in this issue are those focused on the "*purchase intention*" of remanufactured products, and particularly on the purchase intention of consumers of remanufactured products (Wang et al. 2013; Jiménez-Parra et al. 2014; Gaur et al. 2015). Wang et al. (2013) develop a research model that examines the purchase intention of remanufactured automobile spare parts in China, describing how perceived risk, product knowledge and perceived benefits would affect consumers' attitude and intention towards remanufactured products. They employ a model based on the Theory of Planned Behaviour (Ajzen 1985, 1991) to explore the influence of attitude, subjective norm and perceived behavioural control on the purchase intention of remanufactured products. In addition, the model also includes perceived risk, product knowledge and perceived benefit as explanatory variables for purchase attitude and intention. As main results of that research, it can be highlighted that purchase intention is mostly influenced by purchase attitude and there exists a negative correlation between product knowledge and purchase intention, so the more consumers understand remanufacturing processes the less likely they are to purchase remanufactured products. This last surprising finding is explained by Wang et al. (2013) by using the argument that consumers do not trust the remanufacturing process and so they do not perceive the quality equivalent between remanufactured products and original ones. Jiménez-Parra et al. (2014) conduct a research aimed at describing the purchase intention of consumers of remanufactured laptops in Spain, by using a model also based on the Theory of Planned Behaviour. They consider that purchase intention can be influenced by four variables namely attitude towards the purchasing, subjective norm, motivations, and marketing mix variables. Similarly to Wang et al. (2013), they find that a favourable attitude towards remanufactured products positively influences the purchase intention, so it could be assumed that there exists a segment of consumers who are favourably inclined to buying remanufactured products—in that case, remanufactured laptops—and that promotional actions of firms in the sector clearly should be directed at them. Moreover, this study shows that motivations play a significant positive role in the purchase intention. Overall, the respondents showed that price and environmental issues constitute positive motivations for their intention to purchase a remanufactured laptop, as it has also been noted in other studies (Guide and Van Wassenhove 2001; Atasu et al. 2010; Agrawal et al. 2015b). Regarding marketing issue, Jiménez-Parra et al. (2014) point out: "*OEMs and remanufacturers could orient their marketing policies towards actions aimed at identifying consumers with a more favourable attitude towards these products. Their marketing campaigns could be directed not only at these consumers themselves but also at their closest social circles, which we have found to be an important referent in the intention to purchase*". In their study, Gaur et al. (2015) analyse the most relevant drivers of consumer purchase intentions for remanufactured products to the India-born consumers residing in the USA, in order to confirm if being relocated to a society where remanufactured products are fostered can change consumer's perceptions towards them. To this end, the authors develop an active diagram in which the decision-making process of consumers' purchase intentions for remanufactured products is showed. The results suggest that personal factors (i.e., attitudes and beliefs, individual personality and environmental consciousness) have a positive influence on the consumers' decision making and purchase intentions. Furthermore, the findings confirm that contextual factors (i.e., societal norms, price, promotion activities, service quality and brand image) are relevant drivers of consumer behaviour as well. Societal norms, price and service quality have a positive influence on purchase intention for remanufactured products. Additionally, in the same way that in Subramanian and Subramanyam (2012), the authors find that the brand image of the OEM, significantly affect the consumer purchase intention. In this sense, respondents of the Gaur et al.'s study claimed that "*if a good brand offers a remanufactured product, then they would purchase those without any hesitation*". And, similarly to Atasu et al. (2010) and Jiménez-Parra et al. (2014), Gaur et al. (2015) emphasize the importance of promotion activities to enhance consumer WTP remanufactured products and point out the existence of different consumer segments ("green consumers" and "functional conscious consumers") interested in demanding these kinds of products, because they value them equally than original (new) ones. This set of papers is characterised by the use of methodologies and techniques not usual in the context of the research on reverse logistics, such as structural equation models and surveys. Although all these papers make a valuable contribution to this sub-topic of research, some aspects should be considered in the future. We must mention the need to describe the characteristics of the purchase intention in a more general way, and not only for a very specific remanufactured product (laptops, automobile spare parts). In this sense, studies focused on describing the purchase intention of consumers in a wider segment of the market (for example, not just laptops but electronic products) would be welcomed. Moreover, the use of student samples (Jiménez-Parra et al. 2014; Gaur et al. 2015) is arguable, and can be considered as a serious limitation of this kind of studies. Actually, some journals discourage this type of submissions (Bello et al. 2009).

4.4 The Willingness to Pay for Remanufactured Products

Remanufactured products are usually considered as products to be aimed at a segment of consumers with lower WTP, since they value remanufactured products lower than original ones, likely because of the quality perceptions that consumers have about them (Hazen et al. 2012). Consumers' perceptions and "*WTP*" for remanufactured products is an inevitable question when studying product acquisition, remanufacturing, and market competition (Guide and Li 2010). Therefore, this is a relevant issue in the research on CLSC from this "*demand side*". Michaud and Llerena (2011) use experimental auctions to describe consumers' WTP for remanufactured products, finding evidence that consumers tend to value the remanufactured product (in this case, a single-use camera) less than the original one, unless they are informed about their respective environmental impacts. Moreover, the results suggest that these consumers are not willing to pay a premium for the remanufactured product as can be observed for other kinds of products (organic foods, for example). Ovchinnikov (2011) describes how a company makes remanufacturing decisions and provides a study of consumer behaviour based on an estimation of the fraction of consumers that, for a given price difference, would switch from the new to remanufactured product. Some remarkable findings of this paper are that consumers would be more willing to pay for remanufactured products if they had a clear information about terminology used in this market (refurbished, returned, rebuilt, remanufactured, etc.) and about product's history (why was it returned, when, where?). Because of this absence of information, consumers tend to use price as a way to judge product quality, so a low price would be indicating a low product quality, and only a few consumers switch from new to remanufactured product. Essoussi and Linton (2010) analyse the consumers' WTP for recycled products considering seven different types of products (paper, single use cameras, toner cartridges, tyres, auto parts, cell phones, and printers), as well as the switching behaviour of consumers from recycled to new products due to price differences. These authors find that consumers' WTP premium price for recycled products is product specific and there exists a tremendous variation in both relative price and switching range for different types of products. For example, in this case, consumers were more willing to pay premium prices for recycled paper and single use cameras than for toner, cell phones and auto parts. Research papers included in this sub-topic employ methodologies usual in the areas of management and marketing such as surveys, focus group or conjoint analysis. Differences found in the WTP for different products lead us to suggest that more research is needed in order to understand the consumer behaviour of remanufactured products across industries. In addition, issues such as information about recovered products (recycled and remanufactured), the effects of branding and the image of OEM and 3PR on WTP could be elements to consider for further research.

4.5 Consumer Perceptions of Remanufactured Products

Hazen et al. (2012), Subramanian and Subramanyam (2012), Agrawal et al. (2015b), and Wang and Hazen (2015) discuss about different aspects of the "*consumer perceptions of remanufactured products*". For example, in their study, Hazen et al. (2012) conclude that ambiguity inherent in the remanufacturing process (overall age of the product, components used, quality levels, performance, etc.) provokes a consumers' perception of lower quality and reduces their WTP for these items. In a similar way, Subramanian and Subramanyam (2012) evaluate several factors that can explain purchase price differentials between new and remanufactured products, and find evidence that seller reputation and the remanufacturer identity can contribute to reduce the uncertainty about the quality of remanufactured products. Based on a previous research (Wang et al. 2013), Wang and Hazen (2015) further develop the consumer perceptions of remanufactured products. They build a model rooted in Prospect Theory (Kahneman and Tversky 1979; Tversky and Kahneman 1992) in order to analyse how consumer knowledge of remanufactured products—in terms of cost, quality and green attributes—affects Chinese consumers' perception of both risk and value associated with the purchase of these types of products (in this case, remanufactured automobile spare parts). The results confirm that knowledge about product quality has the most effect on consumers' perceptions (risk and value) of remanufactured products. As noted in previous studies (see for example, Michaud and Llerena 2011; Ovchinnikov 2011), *quality* is an important product attribute that consumers bear in mind when they make buying decisions, so they could even reject the remanufactured product if they hesitate about its quality level. In the same vein, both *cost knowledge* and *green knowledge* have significant impacts on consumers' perceptions of remanufactured products so, as Wang et al. (2013) and Jiménez-Parra et al. (2014) suggest, the authors recommend OEMs developing marketing strategies focused on those attributes (lower price and "environmentally friendly") which are the most important ones to consumer segments willing to buy remanufactured products. Conversely, in their study, Agrawal et al. (2015b) investigate how remanufactured products (i.e., consumer's perceptions of these products) and the identity of the remanufacturer (OEM or 3PR) can influence the perceived value of an OEM's new product, by using behavioural experiments. According to the main results, the authors suggest that remanufactured products by the OEM may reduce the perceived value of new products and remanufactured products by the 3PR may increase the perceived value of new products. In line with previous comments, for this sub-topic more research is required to examine how the effects of information, knowledge, and the identity of the remanufacturer (OEM vs. 3PR) influence the consumer perceptions across different brands, product categories, and markets.

Although they may seem many references on this topic, certainly they are just a very little percentage over the total of published work on CLSC, in spite of the fact that some of the most cited authors in this area are claiming for more research in issues as consumer behaviour and the market for remanufactured products (Guide and Van Wassenhove 2009; Atasu et al. 2010; Souza 2013).

Of course, there are still many issues to be analysed—for example, those relating to the potential commercial activities and marketing policies that firms might implement according to the purchasing preferences showed by respondents. Nevertheless, this investigation should be considered as a first step in this process of connecting Marketing with Reverse Logistics, in what we trust will be a fruitful relationship.

5 Conclusions

Reverse logistics is a research topic that has evolved during the last two decades but, at the present time, it can be considered as a well consolidated topic of research, with hundreds of papers published since 1995 in the most prestigious scientific journals. Reverse logistics is not only an interesting issue for researchers, but also for companies and professionals who are considering the recovery of EoU products as a business opportunity, so they are taking into account these activities in their strategic processes of decision making.

Reverse logistics has several implications for the SCM but, probably, the most challenging is related to the design of the RLN. For this reason, a description of the basic activities related to the design process has been examined: collection, inspection, and recovery process. In spite of the relevance of the literature about the design of RLNs, new lines of research are still open.

The interest of academia and professionals about activities related to CLSC, reverse logistics, and remanufacturing has provided a better understanding of the characteristics, the processes and the implications that the recovery of EoU and EoL products have on the business activity. In spite of this fact, there are some concerns that require our attention; for example, those related to the strategic aspects of the CLSC, and particularly those issues related to the potential commercial activities and marketing policies that firms could establish. In this sense, a review of the literature reverse logistics from a "*demand side*" was conducted in order to highlight new opportunities for research in this field. Issues regarding competition in the remanufactured markets, cannibalization, purchase intention of remanufactured products, WTP for them, and consumer perceptions of remanufactured products have been investigated in the last years. However, there still exist relevant questions and issues about the relationships between reverse logistics and the markets that need to be analysed. Quoting Souza (2013, p. 31) '*Research in consumer behavior and the market for remanufactured products is nascent, so a more comprehensive understanding of the market for remanufactured products is needed across different industries*'.

Acknowledgments The authors thank Junta de Extremadura and FEDER for the funding received through grant GR15007 and PRI09A098. In addition, S. Rubio acknowledges the financial support from Junta de Extremadura and FEDER through the grant MOV15B024, and from Ministerio de Economía y Competitividad and FEDER under the project DPI2015-67740-P (MINECO/FEDER).

References

Agrawal, S., Singh, R., & Murtazaet, Q. (2015a). A literature review and perspectives in reverse logistics. *Resources, Conservation and Recycling, 97*, 76–92.

Agrawal, V. V., Atasu, A., & van Ittersum, K. (2015b). Remanufacturing, third-party competition and consumer's perceived value of new products. *Management Science, 61*(1), 60–72.

Ajzen, I. (1985). From intentions to actions: A theory of planed behaviour. In J. Kuhl & J. Beckmann (Eds.), *Action-control: From cognition to behaviour* (pp. 11–39). Heidelberg: Springer.

Azjen, I. (1991). The theory of planned behaviour: Organizational behaviour and human decision. *Processes, 50*(2), 179–211.

Akçali, E., Çetinkaya, S., & Üster, H. (2009). Network design for reverse and closed-loop supply chains: An annotated bibliography of models and solution approaches. *Networks, 53*, 231–248.

Alumur, S. A., Nickel, S., Saldanha-da-Gama, F., & Verter, V. (2012). Multi-period reverse logistics network design. *European Journal of Operational Research, 220*, 67–78.

Aras, N., Boyaci, T., & Verter, V. (2010). Designing the reverse logistics network. In M. E. Ferguson & G. C. Souza (Eds.), *Closed-loop supply chains: New developments to improve*

the sustainability of business practices (pp. 67–97). Boca Raton, Florida: CRC Press, Taylor & Francis Group.
Atasu, A., Guide, V. D. R, Jr., & Van Wassenhove, L. N. (2010). So what if remanufacturing cannibalizes my new product sales? *California Management Review, 52*, 1–21.
Atasu, A., Sarvary, M., & Van Wassenhove, L. N. (2008). Remanufacturing as a marketing strategy. *Management Science, 54*, 1731–1746.
Bello, D., Leung, K., Radebaugh, L., Tung, R. L., & van Witteloostuijn, A. (2009). From the editors: Student samples in international business research. *Journal of International Business Studies, 40*(3), 361–364.
Carrasco, R. (2010). *A management model for closed-loop supply chains of reusable articles: (Doctoral Dissertation)*. Madrid: Universidad Politécnica de Madrid.
Carrasco-Gallego, R., Ponce-Cueto, E., & Dekker, R. (2012). Closed-loop supply chains of reusable articles: a typology grounded on case Studies. *International Journal of Production Research, 50*, 5582–5596.
Corominas, A., Mateo, M., Ribas, I., & Rubio, S. (2015). Methodological elements for the design of supply chains. *International Journal of Production Research, 53*(16), 5017–5030.
De Brito, M.P., & Dekker, R. (2004). A framework for reverse logistics. In R. Dekker, M. Fleischmann, K. Inderfurth, & L. N. Van Wassenhove (Eds.), *Reverse logistics. Quantitative models for closed-loop supply chains* (pp. 3–28). Berlin: Springer.
Debo, L. G., Toktay, L. B., & Van Wassenhove, L. N. (2005). Market segmentation and product technology selection for remanufactured products. *Management Science, 51*, 1193–1205.
Dekker, R., Fleischmann, M., Inderfurth, K., & Van Wassenhove, L. N. (Eds.). (2004). *Reverse logistics. Quantitative models for closed-loop supply chains*. Berlin: Springer.
Dowlatshahi, S. (2005). A strategic framework for the design and implementation of remanufacturing operations in reverse logistics. *International Journal of Production Research, 43*(16), 3455–3480.
Essoussi, L. H., & Linton, J. D. (2010). New or recycled products: How much are consumers willing to pay? *Journal of Consumer Marketing, 27*, 458–468.
Ferguson, M. E., Guide, V. D. R, Jr., & Souza, G. (2006). Supply chain coordination for false failure returns. *Manufacturing Service Operations Management, 8*(4), 376–393.
Ferguson, M. E., & Souza, G. C. (2010). *Closed-loop supply chains: New developments to improve the sustainability of business practices*. Boca Ratón, Florida: CRC Press, Taylor & Francis Group.
Flapper, S. D. P., Van Nunen, J. A. E. E., & Van Wassenhove, L. N. (2005). Introduction. In S. D. P. Flapper, J. A. E. E. Van Nunen, & L. N. Van Wassenhove (Eds.), *Managing closed-loop supply chains* (pp. 3–18). Berlin: Springer.
Fleischmann, M., Bloemhof-Ruwaard, J. M., Dekker, R., van der Laan, E., Van Nunen, J., & Van Wassenhove, L. N. (1997). Quantitative models for reverse logistics: A review. *European Journal of Operational Research, 103*, 1–13.
Fleischmann, M., Krikke, H. R., Dekker, R., & Flapper, S. P. D. (2000). A characterisation of logistics networks for product recovery. *Omega, 28*, 653–666.
Gaur, J., Amini, M., Banerjee, P., & Gupta, R. (2015). Drivers of consumer purchase intentions for remanufactured products. A study of Indian consumers relocated to the USA. *Qualitative Market Research: An International Journal, 18*(1), 30–47.
Ginter, P. M., & Starling, J. M. (1978). Reverse distribution channels for recycling. *California Management Review, 20*(3), 72–81.
Govindan, K., Soleimani, H., & Kannan, D. (2015). Reverse logistics and closed-loop supply chain: A comprehensive review to explore the future. *European Journal of Operational Research, 240*, 603–626.
Guide, V. D. R, Jr., Harrison, T. P., & Van Wassenhove, L. N. (2003). The challenge of closed-loop supply chains. *Interfaces, 336*(6), 3–6.
Guide, V. D. R, Jr., & Li, J. (2010). The potential for cannibalization of new products sales by remanufactured products. *Decision Sciences, 41*, 547–572.

Guide, V. D. R, Jr., & Van Wassenhove, L. N. (2001). Managing product returns for remanufacturing. *Production and Operations Management, 10*(2), 142–155.

Guide, V. D. R, Jr., & Van Wassenhove, L. N. (2009). The evolution of closed-loop supply chain research. *Operations Research, 57*, 10–18.

Guiltinan, J. P., & Nwokoye, N. G. (1974). Reverse channels for recycling: An analysis for alternatives and public policy implications. In R. G. Curhan (Ed.), *New marketing for social and economic progress. Combined proceedings*. Michigan: American Marketing Association.

Hazen, B. T., Overstreet, R. E., Jones-Farmer, L. A., & Field, H. S. (2012). The role of ambiguity tolerance in consumer perception of remanufactured products. *International Journal of Production Economics, 135*, 781–790.

Ilgin, M. A., & Gupta, S. M. (2010). Environmentally conscious manufacturing and product recovery (ECMPRO): A review of the state of the art. *Journal of Environmental Management, 91*, 563–591.

Jiménez-Parra, B., Rubio, S., & Vicente-Molina, M. A. (2014). Key drivers in the behaviour of potential consumers of remanufactured products: A study on laptops in Spain. *Journal of Cleaner Production, 85*, 488–496.

Kahneman, D., & Tversky, A. (1979). Prospect theory: An analysis of decision under risk. *Econometrica, 47*(2), 263–392.

Kopicky, R. J., Berg, M. J., Legg, L., Dasappa, V., & Maggioni, C. (1993). *Reuse and recycling: Reverse logistics opportunities*. Oak Brook, Illinois: Council of Logistics Management.

Lambert, D. M., & Stock, J. R. (1981). *Strategic physical distribution management*. Illinois: Irwin.

Lebreton, B. (2007). *Strategic closed-loop supply chain management*. Berlin: Springer.

Majumder, P., & Groenevelt, H. (2001). Competition in remanufacturing. *Production & Operations Management, 10*, 125–141.

Michaud, C., & Llerena, D. (2011). Green consumer behaviour: An experimental analysis of willingness to pay for remanufactured products. *Business Strategy and the Environment, 20*, 408–420.

Murphy, P. R. (1986). A preliminary study of transportation and warehousing aspects of reverse distribution. *Transportation Journal, 35*(4), 12–21.

Murphy, P. R., & Poist, P. R. (1989). Managing of logistics retromovements: An empirical analysis of literature suggestions. *Transportation Research Forum, 29*(1), 177–184.

Örsdemir, A., Kemahlıoğlu-Ziya, E., & Parlaktürk, A. K. (2014). Competitive quality choice and remanufacturing. *Production & Operations Management, 23*(1), 48–64.

Ovchinnikov, A. (2011). Revenue and cost management for remanufactured products. *Production & Operations Management, 20*, 824–840.

Pohlen, T. L., & Farris, M. T, I. I. (1992). Reverse logistics in plastic recycling. *International Journal of Physical Distribution and Logistics Management, 22*(7), 35–47.

Prahinski, C., & Kocabasoglu, C. (2006). Empirical research opportunities in reverse supply chains. *Omega, 34*, 519–532.

REVLOG (1998). The European working group on reverse logistics. http://www.fbk.eur.nl/OZ/REVLOG

Rogers, D. S., & Tibben-Lembke, R. S. (1999). *Going backwards reverse logistics trends and practices*. Reno: Reverse Logistics Executive Council.

Rubio, S. (2003). El sistema de logística inversa en la empresa. Análisis y aplicaciones (Doctoral Disertation). Universidad de Extremadura.

Rubio, S., Chamorro, A., & Miranda, F. J. (2008). Characteristics of the research on reverse logistics (1995–2005). *International Journal of Production Research, 46*, 1099–1120.

Rubio, S., & Jiménez-Parra, B. (2014). Reverse logistics overview and challenges for supply chain management. *International Journal of Engineering Business and Management, 6*, 1–7.

Savaskan, R. C., Bhattacharya, S., & Van Wassenhove, L. N. (2004). Closed-loop supply chain models with product remanufacturing. *Management Science, 50*, 239–252.

Souza, G. (2013). Closed-loop supply chains: A critical review, and future research. *Decisions Sciences, 44*, 7–38.

Stock, J. R. (1992). *Reverse logistics*. Oak Brook, Illinois: Council of Logistics Management.

Subramanian, R., & Subramanyam, R. (2012). Key factors in the market for remanufactured products. *Manufacturing & Service Operations Management, 14*, 315–326.

Thierry, M., Salomon, M., Van Nunen, J., & Van Wassenhove, L. N. (1995). Strategic issues in product recovery management. *California Management Review, 37*, 114–135.

Toktay, L. B., Wein, L., & Zenios, S. (2000). Inventory management of remanufacturable products. *Management Science, 46*(11), 1412–1426.

Tversky, A., & Kahneman, D. (1992). Advances in prospect theory: Cumulative representation of uncertainty. *Journal of Risk and Uncertainty, 5*(4), 297–323.

Verstrepen, S., Cruijssen, F., de Brito, M. P., & Dullaert, W. (2007). An exploratory analysis of reverse logistics in Flanders. *European Journal of Transport and Infrastructure Research, 7*, 301–316.

Wang, Y., & Hazen, B. T. (2015). Consumer product knowledge and intention to purchase remanufactured products. *International Journal of Production Economics* (in press).

Wang, Y., Wiegerinck, V., Krikke, H., & Zhang, H. (2013). Understanding the purchase intention towards remanufactured product in closed-loop supply chains: An empirical study in China. *International Journal of Physical Production and Logistics Management, 43*(10), 866–888.

Yang, J., Golany, B., & Yu, G. (2005). A concave-cost production planning problem with remanufacturing options. *Naval Research Logistics, 52*(5), 443–458.

Zhou, Z.-B., Wang, J.-J., Deng, G.-S., & Chen, H. (2016). Third-party remanufacturing mode selection: Outsourcing or authorization? *Transportation Research Part E, 87*, 1–19.

Optimization Concepts—I: Introductory Level

Miguel Casquilho and João Luís de Miranda

Abstract Optimization subjects are addressed from both an introductory and an advanced standpoint. Theoretical subjects and practical issues are focused, conjugating Optimization basics with the implementation of useful tools and SC (supply chain) models. By now, in the *Introductory* section, Linear Programming, Integer Programming, related models, and others of interest are treated. In another chapter about these subjects in this handbook, more advanced related topics are addressed.

Keywords Optimization · Supply chain · Linear Programming

1 Introduction

Optimization is the act or methodology of driving something, such as a system, as near as possible to full functionality or effectiveness, the word having appeared one and a half century ago. Nowadays, the concept is essential, in a global context, because even a small improvement in a mathematical, economical or industrial solution can be decisive in comparison with other, nonoptimal solutions.

A crucial contribution to optimization was the discovery of a computationally simple method to solve "linear programming" problems, in the 1940s, by George Dantzig, the so-called *simplex method.* This method, which is important in the

M. Casquilho (✉)
Department of Chemical Engineering, Instituto Superior Técnico,
Ave. Rovisco Pais IST, 1049-001 Lisbon, Portugal
e-mail: mcasquilho@tecnico.ulisboa.pt

J.L. de Miranda
Departamento de Tecnologias e Design, Escola Superior de Tecnologia e Gestão,
Instituto Politécnico de Portalegre, 7300-110 Portalegre, Portugal
e-mail: jlmiranda@estgp.pt

M. Casquilho · J.L. de Miranda
CERENA—Centro de Recursos Naturais e Ambiente, Instituto Superior Técnico,
University of Lisbon, Ave. Rovisco Pais, IST, 1049-001 Lisbon, Portugal

A.P. Barbosa-Póvoa et al. (eds.), *Optimization and Decision Support Systems for Supply Chains*, Lecture Notes in Logistics,
DOI 10.1007/978-3-319-42421-7_5

Supply Chain problems, has since been very much used and studied, and important extensions have been obtained, some maintaining the efficiency of the original algorithm (as the Transportation Problem, the Transshipment Problem, the Assignment Problem), some addressing other, more complex situations, namely Integer Programming, which uses Linear Programming as a tool included in other techniques, of which the branch-and-bound is explained in this text. (Source computer code of some of these subjects can be obtained on request.)

2 Linear Programming (LP)

"The Best of the 20th Century: Editors Name Top 10 Algorithms" (Cipra 2000)

1947: George Dantzig, at the RAND Corporation, has created the simplex method for linear programming. In terms of widespread application, Dantzig's algorithm is one of the most successful of all times: linear programming dominates the world of industry, where economic survival depends on the ability to optimize within budgetary and many other constraints. (Of course, the "real" problems of industry are often nonlinear, and the use of linear programming is sometimes dictated by the computational budget.) The simplex method appears an elegant way of arriving at optimal answers. Although theoretically susceptible to exponential delays, the algorithm in practice is highly efficient—which in itself says something interesting about the nature of computation.

In the following, some examples are given, the theory of which can be found in several classical books, such as Zionts (1974), Bronson (2010), Hillier and Lieberman (2009). An intuitive algebraic approach is given below to solve linear programming problems. It is meant to show that the method is iterative, as is readily revealed.

$$\begin{array}{rllll} [\max] z = & 0.56x_1 & +0.42x_2 & & \\ \text{s. to} & x_1 & +2x_2 & \leq & 240 \\ & 1.5x_1 & +x_2 & \leq & 180 \\ & x_1 & & \leq & 110 \end{array} \tag{1}$$

$$\mathbf{A} \quad \begin{array}{rllllll} [\max] z = & 0.56x_1 & +0.42x_2 & & & & \\ & x_1 & +2x_2 & +\{x_3\} & & & = 240 \\ & 1.5x_1 & +x_2 & & +\{x_4\} & & = 180 \\ & x_1 & & & & +\{x_5\} & = 110 \end{array} \tag{2}$$

This has (always) an obvious, sure solution. Let

$$x_1, x_2 = 0 \tag{3}$$

Then

$$\begin{bmatrix} x_3 \\ x_4 \\ x_5 \end{bmatrix} = \begin{bmatrix} 240 \\ 180 \\ 110 \end{bmatrix} \tag{4}$$

$$z = \begin{bmatrix} 0 & 0 & 0 \end{bmatrix} \begin{bmatrix} 240 \\ 180 \\ 110 \end{bmatrix} = 0 \tag{5}$$

Is this optimal? How to improve it?

There does not appear (Hillier and Lieberman 2005) to exist a systematic way of setting *all* the nonbasic variables, i.e., those that were made zero, simultaneously to optimal values—hence, an *iterative* method should be followed.

Choose the variable that increases the objective function *most* per unit (this choice is somehow arbitrary), in the example, x_1, because its coefficient (0.56) is the largest.

According to the constraints, x_1 can be increased till:

$$\mathbf{B} \quad \begin{array}{ll} x_1 & = 240 \\ 1.5x_1 & = 180 \\ x_1 & = 110 \end{array} \quad \rightarrow \quad \begin{array}{ll} x_1 & = 240 \\ x_1 & = 120 \\ x_1 & = 110 \end{array} \tag{6}$$

The *third* equation in (2) leads to $x_1 = 110$ and $x_5 = 0$. This choice comes from the fact that it determines the smallest, most stringent limit. If this limit were exceeded, at least one other variable would become negative. The variable x_1 will be the *entering* variable and x_5 the *leaving* variable:

$$\mathbf{C} \quad x_1 = 110 - x_5 \tag{7}$$

Substituting for x_1 everywhere (except in "its own" constraint, i.e., the one that led to its choice), we have

$$\begin{array}{lllllll} [\max]z = & 0.56(110 - x_5) & +0.42x_2 & & & & \\ & (110 - x_5) & +2x_2 & +x_3 & & & = 240 \\ & 1.5(110 - x_5) & +x_2 & & +x_4 & & = 180 \\ & x_1 & & & & +x_5 & = 110 \end{array} \tag{8}$$

$$\mathbf{A} \quad \begin{array}{lllllll} [\max]z = & & 0.42x_2 & & & -0.56x_5 & +61.6 \\ & & +2x_2 & +\{x_3\} & & -x_5 & = 130 \\ & & x_2 & & +\{x_4\} & -1.5x_5 & = 15 \\ & \{x_1\} & & & & +x_5 & = 110 \end{array} \tag{9}$$

which is equivalent to Eq. (2).

We now have a **new** (equivalent) LP problem, **to be treated as the original was:** the essence of the simplex method has just been found! The process can continue *iteratively*, until there is no variable leading to improvement in the objective function. In sum:

A. In the system of equations, find the identity matrix (immediate solution).
B. Search for an *entering* variable (or finish).
C. Consequently, find a *leaving* variable (if wrongly chosen, negative values will appear).

3 Transportation Problem and special cases

In the supply chain environment, several problems related to transportation and others apparently unrelated can be formulated and solved by the technique used for the typical *transportation problem*, frequently simply denoted by the initials TP. Besides the TP, we shall address: the (simple) *production scheduling*; the *transshipment problem*; and the *assignment problem* (AP). These problems can be solved by their own algorithms: the TP, the production scheduling and the transshipment, by the "stepping-stone" method; and the AP by the Hungarian method. As all these problems are particular cases of Linear Programming (LP), the problems will be presented and then formulated as LP problems. Indeed, with the current availability of high quality LP software, namely, the suggested IBM ILOG CPLEX (2015a, b, c), it looks unnecessary to go into the details of those other methods. Otherwise, a negative remark about the Hungarian method to solve the AP is that, although it is easy and elegant to be done by hand, it is quite hard to program, with no readily available source code in the literature.

The general goal is to "transport" (whatever that may be) goods to the customers at minimum global cost of transportation, according to the unit costs of transportation (certainly dependent on distance, etc.) from the sources to the destinations. The problems mentioned are dealt with in the following sections, mainly based on examples.

4 The Transportation Problem

The Transportation Problem (TP) arises from the need of programming the optimal distribution of a *single* product from given sources (supply) to given destinations (demand).

The product is available in m sources, with known quantities (also said *capacities*), a_i, $i = 1 .. m$ (the two dots denoting a range, as used in various computer languages, avoiding the vague 1, …, m), and is needed in n destinations, with known quantities (or capacities), b_j, $j = 1 .. n$, and will be sent directly from the sources to the destinations at unit costs, c_{ij}, all these values being the known data.

The objective is to find the quantities to be transported, x_{ij}, at minimum global cost, usually in a given time period, such as a week. The problem can thus be formulated according to the model of Eq. (10), where obviously the sum of the a's must equal that of the b's, without which the problem would be infeasible.

$$[\min]z = \sum_{i=1}^{n}\sum_{j=1}^{n} c_{ij}x_{ij}$$

Subject to

$$\sum_{j=1}^{n} x_{ij} = a_i \quad i = 1\ldots m \tag{10}$$

$$\sum_{i=1}^{m} x_{ij} = b_j \quad j = 1\ldots n$$

$$x_{ij} \geq 0 \quad \forall i,j$$

The physical (typically integer) units of x, a and b can be, say, kg (or m^3, bags, etc.), and c in $/kg (with $ a generic money unit, such as dollar, euro). The scheme in Fig. 1 makes the problem clear.

The model in Eq. (10) is, of course, in all its components (including the last one, of non-negativity of the variables), an instance of Linear Programming (LP). Our notation "[max]" ([min], [opt]) means that the maximum of *both* sides is required, and not that the maximum of z, the objective function, is equal to the right-hand side (otherwise even not yet known).

It is remarkable that the TP can be envisaged as an integer programming problem. The x's will always be, namely in the optimum, multiples of the greatest common divisor of the set of a's and b's. So, if these are integers (as is usual), the

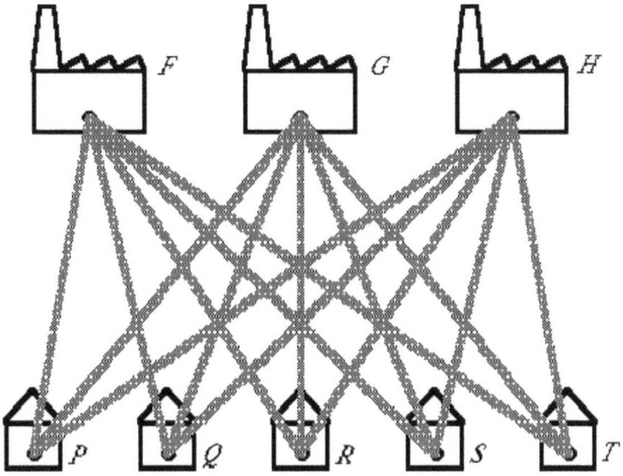

Fig. 1 Transportation problem: from 3 factories to 5 warehouses

x's will be integers; if, e.g., these are multiples of 7, so will they be, etc. If the problem is stated with "decimals", as 4.7, the x's will be multiples of 0.1, so, with appropriate multiplication by a suitable constant, the results will be integer.

Any problem having the above structure can be considered a TP, whatever may be the subject under analysis.

Example

A company, as in Fig. 1, has 3 production centres, factories F, G and H, in given locations (different or even coincident) with production capacities of 100, 120, and 120 ton (per day), respectively, of a certain (single) product with which it must supply 5 warehouses, P, Q, R, S, and T, needing 40, 50, 70, 90, and 90 ton (per day), respectively. The unit costs of transportation, the matrix **C**, are those in Table 1. Determine the most economical (cheapest) transportation plan, matrix **X**.

Resolution

Introduce the transportation matrix, **X**, in Eq. (11), the values of whose elements must be found. (Notice that z, in Eq. (10), does not result from a typical product of matrices!)

$$\mathbf{X} = \begin{bmatrix} x_{11} & x_{12} & x_{13} & x_{14} & x_{15} \\ x_{21} & x_{22} & x_{23} & x_{24} & x_{25} \\ x_{31} & x_{32} & x_{33} & x_{34} & x_{35} \end{bmatrix} \quad (11)$$

The problem could be solved by the adequate "stepping-stone" method (which is simple and very efficient), but its formulation leads directly to the LP in Eq. (12).

$$\begin{aligned} [\min] z = {} & c_{11}x_{11} + c_{12}x_{12} + c_{13}x_{13} + c_{14}x_{14} + c_{15}x_{15} \\ & + c_{21}x_{21} + \cdots + c_{25}x_{25} \\ & + c_{31}x_{31} + \cdots + c_{35}x_{35} \end{aligned}$$

Table 1 Costs of transportation ($/ton) from the factories to the warehouses

	P	Q	R	S	T	
F	4	1	2	6	9	100
G	6	4	3	5	7	120
H	5	2	6	4	8	120
	40	50	70	90	90	

Subject to

$$
\begin{aligned}
x_{11} + x_{12} + x_{13} + x_{14} + x_{15} &= a_1 \\
x_{21} + x_{22} + x_{23} + x_{24} + x_{25} &= a_2 \\
x_{31} + x_{32} + x_{33} + x_{34} + x_{35} &= a_3
\end{aligned}
\qquad (12)
$$

$$
\begin{aligned}
x_{11} + x_{21} + x_{31} &= b_1 \\
x_{12} + x_{22} + x_{32} &= b_2 \\
&\ldots \\
x_{15} + x_{25} + x_{35} &= b_5
\end{aligned}
$$

A TP has really $m + n - 1$ *independent* constraints (not $m + n$), as (any) one of the constraints shown above is superfluous (dependent). This is due to the sheer nature of the TP, in which total supply must equal total demand. If, as happens in several circumstances, supply and demand are not equal, one (no need for more) fictitious entity (source or destination) is introduced, as detailed below.

The fact that the current solvers can be embedded in commercial worksheets, means they natively accept the TP in tabular form and the constraints in Eq. (12) are readily available for solution. The solution is given in Table 2, with a minimum global cost of $z^* = 1400$ \$ (per day, the period considered). In this particular problem, it happens that there are two (i.e., multiple) solutions, the other differing in $x_{11} = 10$, $x_{12} = 50$, $x_{31} = 30$ and $x_{32} = 0$.

The TP is, naturally, "balanced", i.e., the total supply is equal to the total demand. As mentioned, in the cases where there is excess *supply*, the problem can be readily converted to a TP by creating (at cost 0) one *fictitious destination*; or if there is excess *demand* (insufficient supply), one *fictitious source*. So, product could, respectively, be left "at home" or, possibly, bought from some competitor to guarantee the supply to the customer.

Table 2 Quantities to be transported (ton) from the factories to the warehouses

	P	Q	R	S	T	
F	40	20	40			100
G			30		90	120
H		30		90		120
	40	50	70	90	90	

5 The Production Scheduling

The (simple) "production scheduling" problem will be presented through an example akin to many in the literature, such as Hillier and Lieberman (2006, pp. 330–331).

Example
See Table 3.

Resolution
The "transportation" in the production scheduling is not in space, but in time, between months in this example. Assuming the basic TP problem where backorders are not considered production of a certain month is not used to supply the previous month, thus the corresponding unit costs should be prohibited, making them "very large", say, M (the classical "big M"), infinity, or, for computing purposes, sufficiently large (in this problem, e.g., 100 will be enough). Using data in Tables 3 and 4 is obtained, by adding the storage costs and introducing a fictitious fifth month for balancing.

In order to define a sufficiently large M, try some "reasonably" large value, i.e., at least large compared to the other cost values in the problem. (The naïve choice of the *greatest* number in the computer is not valid, because of probable overflow.) If this value is effective in the solution (prohibiting the related x's), then it is a good choice, but, if it is not effective (too small), try a greater new value. If the value is "never" sufficiently large, then, the problem has no physical solution (is *impossible*), although it always has a mathematical one.

The solution has a (minimum) global cost of 82.7 $ with the production schedule given in Table 5.

Table 3 Production scheduling data for Manufacture Co.

Month	Scheduled installations	Max. production	Unit production cost	Unit storage cost
1	15	25	1.08	0.015
2	15	35	1.11	0.015
3	25	30	1.10	0.015
4	20	10	1.13	0.015

Table 4 TP-like data for the Northern Airplane Co. problem

Month	1	2	3	4	(5)	Supply
1	1080	1095	1110	1125	0	25
2	M	1110	1125	1140	0	35
3	M	M	1100	1115	0	30
4	M	M	M	1130	0	10
Demand	15	15	25	20	25	**(100)**

Table 5 Production schedule for the Northern Airplane Co. problem

Month	1	2	3	4	(5)	Supply
1	**15**	**10**	**5**	**0**	0	25
2	–	**5**	**0**	**0**	30	35
3	–	–	**20**	**10**	0	30
4	–	–	–	**10**	0	10
Demand	15	15	25	20	25	**(100)**

6 The Transshipment Problem and the Assignment Problem

The transshipment problem and the assignment problem (AP) can be considered problems reducible to TP's. The transshipment is typically treated like a TP, whereas the AP has the Hungarian algorithm, which is very efficient. Notwithstanding, this algorithm will not be presented, as the AP is a particular LP and is appropriately and easily solved by the usual LP software.

The method to reduce a transshipment problem to a TP is simply to consider that the transshipment points are supply points or demand points or both, by inserting them on the supply side or the demand side (or both). Upon inserting these transshipments as referred, each individual capacity must be "corrected" by adding to it the original global capacity.

The AP is a particular case of the TP, having a square cost matrix and being soluble by considering all the values of supply as 1 and all the values of demand also as 1.

Example, Transshipment

The example is represented in Fig. 2 and is akin to problems from the literature.

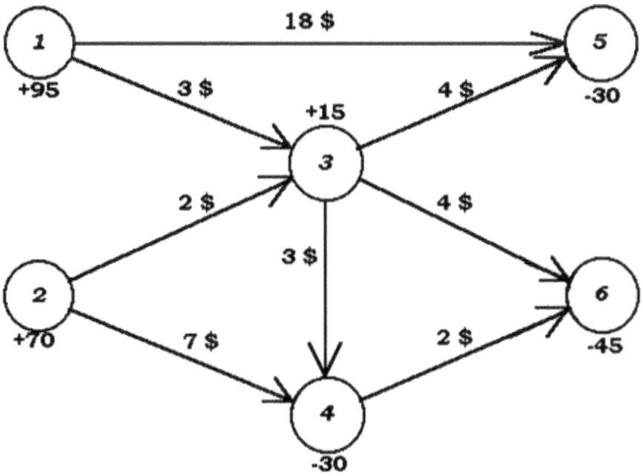

Fig. 2 Transshipment problem: sources (1 and 2), destinations (5 and 6), and transshipment points (3 and 4)

Resolution

In order to convert the transshipment to a TP, identify: (a) every pure supply point (producing only), usually labelled with a positive quantity, such as Point 1 with +95 units; (b) every pure demand point (receiving only), usually labelled with a negative quantity, such as Point 5 with −30 units; and (c) every mixed point (producing or receiving), labelled with a positive (if net producer) or negative (if net receiver) quantity, such as Point 3 with +15 units. Make the original TP balanced, which results here in a dummy destination (Point 7), and register the original capacity of the TP, Q (here $Q = 180$).

The cost matrix becomes the one in Table 6. (The number of times Q is inserted on the supply side and on the demand side is, of course, the same, thus maintaining the equilibrium necessary for a TP.) The solution is in Table 7.

So: from Point 1, 20 units go to Point 3, and 75 stay home; from Point 2, 70 units go to Point 3; from Point 3, 30 go to Point 4, etc.; and Point 4 just receives 30 (from Point 3), with its quantity (equal to Q) meaning it was not used as a transshipment point.

Example, Assignment

Suppose that n tasks are to be accomplished by n workers and the workers have the abilities for every task as given in Table 8.

Table 6 Cost matrix for the transshipment problem

	3	4	5	6	(7)	Supply
1	3	M	18	M	0	95
2	2	7	M	M	0	70
3	0	3	4	4	0	$15 + Q$
4	M	0	M	2	0	Q
Demand	Q	$30 + Q$	30	45	75	$(180 + 2Q)$

Table 7 Solution to the transshipment problem

	3	4	5	6	(7)	Supply
1	20	–	0	–	75	95
2	70	0	–	–	0	70
3	90	30	30	45	0	$15 + Q$
4	–	180	–	0	0	Q
Demand	Q	$30 + Q$	30	45	75	$(180 + 2Q)$

Table 8 Ability of each worker for each task

	1	*2*	*3*	*4*
1	15	16	14	14
2	14	14	13	15
3	13	15	13	14
4	15	16	14	14

Table 9 Assignments (solution)

	1	*2*	*3*	*4*
1	1	0	0	0
2	0	0	0	1
3	0	0	1	0
4	0	1	0	0

These "positive" abilities are preferably converted to "costs", replacing each element by, e.g., its difference to their maximum (here, 16). After that, solve the AP as a TP with each supply equal to 1 and each demand also equal to 1. The solution to this example is in Table 9.

So, Worker 1 does Task 1, Worker 2 does Task 4, etc. (i.e., 1-1, 2-4, 3-3, 4-2), at a minimum global cost of 5 cost units. This particular problem has multiple solutions, another being 1-2, 2-4, 3-3, 4-1.

As this problem has, obviously, always n elements of value 1 (the assignments) and the remaining $n^2 - n$ of value zero, its optimum solution is very *degenerate* if, as was done here, it is considered a TP (degenerate in relation to the $m + n - 1$ possible positive cells in a TP). This is an argument in favour of the Hungarian method, but the strength of that method is not significant when common software is used. (The algorithm of Jonker and Volgenant would be preferable, if the AP itself is under study.)

In the supply chain, several problems related to transportation can be formulated and solved by the technique used for the typical transportation problem (TP), with its own very efficient algorithm (stepping-stone): the (simple) production scheduling; the transshipment problem; and the assignment problem (AP). The AP can be solved by its own algorithm, but the availability of software for Linear Programming makes it practical to solve them as TP's, after convenient simple conversions.

7 Integer Programming by Branch-and-Bound

An example from Ecker and Kupferschmid (1988, p. 217 ff) is used, with modifications and a different, more systematic path to the solution. Using, for example, the simple software Lindo (2013) semi-manually, or CPLEX, the successive solutions can be obtained, to observe the Branch-and-Bound methodology. Of course, these software tools or any other adequate to the problem (possibly) use this technique without necessarily showing the path to the solution.

The problem, shown in (MILP.1), will be to maximize the objective function, z, subject to: three constraints, with the typical constraint of nonnegative independent variables; and being integers. It is this last constraint that makes this a problem of *integer programming* or *mixed integer programming* in the cases where some variables are continuous. Usually, this subject, which is related to Linear Programming, is denoted by MILP, "mixed integer linear programming".

$$\begin{array}{rrlrlrlcr} [\max]z = & -3 & x_1 & +7 & x_2 & +12 & x_3 & & \\ & -3 & x_1 & +6 & x_2 & +8 & x_3 & \leq & 12 \\ \text{subject to} & +6 & x_1 & -3 & x_2 & +7 & x_3 & \leq & 8 \\ & -6 & x_1 & +3 & x_2 & +3 & x_3 & \leq & 5 \\ & & \multicolumn{7}{l}{x_1, x_2, x_3 \text{ nonnegative integers}} \end{array} \qquad \text{(MILP.1)}$$

The resolution of a MILP inherits ***no ease*** from the Linear Programming proper. The typical way to solve a MILP is through the technique of *branch-and-bound*, as explained based on this example.

Begin by solving the MILP as a simple LP. If the solution happens to be integer—which is, of course, rare—, it is the optimum sought. The solution to the original problem, say, P0, is

$$X = [0 \quad 0.303 \quad 1.272] \qquad \text{(MILP.2)}$$

$$z = 17.39$$

The solution, i.e., X, in Eq. (MILP.2) is not integer, so the problem has not been solved. Now, to solve it, let us introduce the B&B technique.

(a) Select any non-integer variable, x_k (prefer the "least integer", i.e., values closer to halves, and ignore draws).
(b) Replace the current problem by two new problems:
   (i) Current problem augmented with constraint $x_k \leq \lfloor x_k \rfloor$, and
   (ii) Current problem augmented with constraint $x_k \geq \lceil x_k \rceil$.
(c) Solve the two problems and repeat, if necessary.

The notation $\lfloor \rfloor$ indicates the 'floor' and $\lceil \rceil$ the 'ceiling' functions (the Iverson notation). So, the original problem will be successively branched: this will occur in an unpredictable way, giving rise to possibly many problems; and also, each time there is branching, the new level will be more difficult than the previous one (one more constraint). Remember that the difficulty (computational complexity) of an LP problem depends essentially on its number of constraints (not its number of variables), which leads to about 1.5 m iterations (some authors preferring 2–3 m), with m the number of constraints. (There are ways, beyond the context of this text, to attenuate this growing difficulty, such as solving dual problems. Also, if there are binary variables, branching leads to just fixing in the simple values of 0 or 1.)

In the example, branching will thus be around the variable $x_2 = 0.303$. The two new problems will be the original problem augmented with one of the two constraints, having now 4 constraints, giving, respectively,

P1	P2
Original problem	Original problem
$x_2 \leq 0$	$x_2 \geq 1$

The solutions to the problems are

$$\text{P1} \;\; X = [0 \quad 0 \quad 1.143] \qquad \text{(MILP.3)}$$

$$z = 13.7$$

and

$$\text{P2} \;\; X = [0.667 \quad 1 \quad 1] \qquad \text{(MILP.4)}$$

$$z = 17$$

In the process of choosing the sub-problem to branch on, it is impossible to predict the best choice. So, select the sub-problem with the best (i.e., most promising) value of the objective function, z. In the example, maybe the process will end up getting an integer solution with $z > z_1$—for instance, $z^* = 14$—, permitting to avoid to explore its branches, so P1 will not be chosen. This avoidance is the advantageous feature of the B&B.

Neither P1 or P2 problems have an integer solution. The more promising is P2, because it presents a greater z, so it is chosen, giving P3 and P4. Branching will be around $x_1 = 0.667$. The two new problems will be the current problem augmented with one of the constraints, having now 5 constraints, giving, respectively,

P3 (from P2)	P4 (from P2)
Original problem	Original problem
$x_2 \geq 1$	$x_2 \geq 1$
$x_1 \leq 0$	$x_1 \geq 1$

The solutions to the problems are

$$\text{P3} \;\; X = 0 \quad 1 \quad 0.667 \qquad \text{(MILP.5)}$$

with

$$z = 15$$

and

$$\text{P4} \;\; X = 1 \quad 1.348485 \quad 0.863636 \qquad \text{(MILP.6)}$$

$$z = 16.8$$

The active problems are now P1 (z = 13.7), P3 (z = 15) and P4 (z = 16.8). Problem P2 (replaced by P3 and P4 has been *fathomed*, i.e., discarded after evaluation. So, P4 will be chosen to branch from, giving P5 $(x_2 \leq 1)$ and P6 $(x_2 \geq 2)$.

The B&B technique proceeds in this way, generating a sequence of solutions The path to the solution (apart from the change of 'min' to 'max of symmetrical') is not coincident with the one given in the example cited in Ecker and Kupferschmid (1988). (In the 3rd row of solutions, the rightmost, 4th, solution, "Infeasible", is possibly wrong, although without further influence.) The integer solution is z^* = 15, with X^* = [2; 3; 0], where of course the term "integer" relates to the values of X^*.

Acknowledgments The Authors thank the College of Technology and Management at the Portalegre Polytechnics Institute (ESTG/IPP) and Instituto Superior Técnico (IST). These works are partially developed at the Centre for Chemical Processes (CPQ/IST) with the support of FCT project PEst-OE/EQB/UI0088, and other developments at CERENA/IST with the support of FCT project UID/ECI/04028/2013.

References

Bronson, R., Naadimuthu, G. (2010). Schaum's outline of theory and problems of Operations Research. New York, NY, USA: McGraw-Hill.

Cipra, B. A. (2000). The best of the 20th century: Editors name Top 10 Algorithms. In J. G. Ecker & M. Kupferschmid (Eds.) (1988). *Introduction to operations research*. New York, NY, USA: Wiley. ISBN 0-471-63362-3. *SIAM News*, *33*(4).

CPLEX. (2012). IBM ILOG CPLEX optimizer.

Ecker, J. G., & Kupferschmid, M. (1988). *Introduction to operations research*. Chichester, UK: Wiley.

Hillier, F. S., & Lieberman, G. J. (2005). *Introduction to operations research*. New York, NY, USA: McGraw-Hill.

Hillier, F. S., & Lieberman, G. J. (2006). *Introduction to operations research*. New York, NY, USA: McGraw-Hill.

Hillier, F. S., & Lieberman, G. J. (2009). *Introduction to operations research*. New York, NY, USA: McGraw-Hill.

IBM ILOG CPLEX Optimization Studio OPL Language User's Manual. (2015a). © Copyright IBM Corp., 1987. http://www-01.ibm.com/support/knowledgecenter/SSSA5P_12.6.2/ilog.odms.studio.help/pdf/opl_languser.pdf?lang=en

IBM ILOG CPLEX Optimization Studio CPLEX User's Manual. (2015b). © Copyright IBM Corporation 1987. http://www-01.ibm.com/support/knowledgecenter/SSSA5P_12.6.2/ilog.odms.studio.help/pdf/usrcplex.pdf?lang=en

IBM ILOG CPLEX Optimization Studio CP Optimizer User's Manual. (2015c). © Copyright IBM Corporation 1987. http://www-01.ibm.com/support/knowledgecenter/SSSA5P_12.6.2/ilog.odms.studio.help/pdf/usrcpoptimizer.pdf?lang=en

Lindo. (2013). Lindo Systems, Inc., http://www.lindo.com/
Taha, H. (1992). *Operations research: An introduction*. New York, NY, USA: MacMillan Publishing Company.
Zionts, S. (1974). *Linear and integer programming*. Englewood Cliffs, NJ, USA: Prentice-Hall. ISBN 0-13-536763-8.

Optimization Concepts: II—A More Advanced Level

João Luís de Miranda and Miguel Casquilho

Abstract Optimization subjects are addressed from both an introductory and an advanced standpoint. Theoretical subjects and practical issues are focused, conjugating Optimization basics with the implementation of useful tools and supply chain (SC) models. In the prior *Introductory* chapter on Optimization concepts, Linear Programming, Integer Programming, related models, and other basic notions were treated. Here, the *More Advanced* chapter is directed to Robust Optimization, complex scheduling and planning applications, thus the reading of the prior *Introductory* chapter is recommended. Through a generalization approach, scheduling and planning models are enlarged from deterministic to stochastic frameworks and robustness is promoted: model robustness, by reducing the statistical measures of solutions variability; and solutions robustness, by reducing the capacity slackness, the non-used capacity of chemical processes that would imply larger investment costs.

Keywords Robust Optimization · Batch scheduling · Process planning · Models generalization

---

J.L. de Miranda (✉)
Departamento de Tecnologias e Design, Escola Superior de Tecnologia e Gestão, Instituto Politécnico de Portalegre, 7300-110 Portalegre, Portugal
e-mail: jlmiranda@estgp.pt

M. Casquilho
Department of Chemical Engineering, Instituto Superior Técnico, Ave. Rovisco Pais, IST, 1049–001 Lisbon, Portugal
e-mail: mcasquilho@tecnico.ulisboa.pt

J.L. de Miranda · M. Casquilho
CERENA—Centro de Recursos Naturais e Ambiente, Instituto Superior Técnico, University of Lisbon, Ave. Rovisco Pais, IST, 1049-001 Lisbon, Portugal

A.P. Barbosa-Póvoa et al. (eds.), *Optimization and Decision Support Systems for Supply Chains*, Lecture Notes in Logistics,
DOI 10.1007/978-3-319-42421-7_6

1 Introduction

Generalization approaches occur widely in Mathematics fields, namely, the enlargement of number sets, from the set of natural until the set of complex numbers, considering the set of integer, fractional, and real numbers. These successive enlargements occur also in the power operations, when the powers of natural, integer, fractional, real, and complex numbers are addressed. Or in certain functions, such as the Permutation function can be considered just a special case of Factorial function, the "complete factorial", and the Factorial function just a restriction to non-negative integers of the Gamma function.

But generalizations occur also in Mathematical Programming (MP), since Linear Algebra solutions are considered in the Linear Programming (LP) feasible solutions space, the methods for Integer LP are using constrained-LP, deterministic LP are enlarged to Stochastic LP (SLP) and then to Robust Optimization (RO) models (Miranda and Nagy 2011). In fact, the analysis of the evolution of MP models, with quantitative and qualitative improvements, show successively extended models. Quantitative improvements are achieved when the objective function and/or space of admissible solutions are successively enlarged, and vice versa.

The study of several industry-based SC cases addressing petroleum refineries, fertilizers, pharmaceutics, chemical specialties, and paper production was developed in a first phase (Miranda 2007). The focus was on Mixed Integer Linear Programming (MILP) models that aim at chemical processes planning (PP), that is, long range investment planning.

Notice that in the chemical industry:

- Dedicated chemical processes, on one hand, use fixed proportions of products in all the time periods; for instance, in the Kellogg's process for ammonia synthesis and in the electrochemical production of caustic soda;
- On the other hand, the flexible chemical processes that occur in petroleum refineries use different products along the various time periods.
- Process flexibility may occur in raw materials, namely, some refinery processes are suitable for several types of crude oil; and process flexibility may also occur in products; for instance, when producing several types of paper, considering different surface properties, various densities and colors.

The successive enlargement of models associated with the generalization approach was observed, with PP models treating dedicated processes, then flexible processes, and flexible processes through "production schemes". Furthermore, robust planning with dedicated processes, and SC production planning with flexible processes were also treated, but not simultaneously in the same PP model.

Another focus on the industry cases was on MILP models aiming at the design and scheduling of chemical batch processes (here on this text: *Batch*, a *flowshop*-type problem), considering the production of tires, biotechnology, food industry, chemical specialties, pharmaceutics. The *flowshop* problem is a very well-known type of problem and addresses N tasks to be sequentially performed in M stages, assuming (Garey et al. 1976): (i) one machine is available in each stage; (ii) one task is performed in one machine at each allocation, without preemption; and (iii) one machine performs only one task at each time. Beyond the wide number of applications in manufacturing, and due to the high cost of equipment in chemical batch processes, either the maximization of equipment utilization or the make-span minimization are usually considered.

Like in the former PP cases, the evolution of *Batch* models is observed in two senses: (i) successively, more realistic formulations, integrating additional degrees of freedom and uncertainty treatment, which are making resolution more difficult; and (ii) usually, resolution procedures are considered at the very beginning of the modeling phase, and some issues of networks resolution, heuristics, and approximation schemes, were introduced. Due to its inherent difficulty, the *Batch* model at hand was treated only in a deterministic framework, and further developments except this generalization approach are not at our knowledge.

In order to obtain quantitative and qualitative improvements, then, a PP model and a *Batch* model are selected and generalized, aiming at the treatment of uncertainty and promotion of robustness. Simultaneously, these models also address the chemical processes designing and sizing, the material fluxes within such processes, and finally materials to purchase and products to sell.

2 Robust Model for Flexible Processes Planning

As an overview of the PP problem and considering the evolution of PP models during about a decade, a capacity expansion sub-problem was considered and, simultaneously, uncertainty is treated (Sahinidis et al. 1989), flexibility attributes are included (Norton and Grossmann 1994), and robustness is promoted (Bok et al. 1998).

The *Capacity Expansion* problem, in general, addresses the expansion of production capacities when products demand are expected to rise significantly. It considers economies of scale by modeling fixed and variable costs and, in particular, obsolescence (e.g., electronics manufacturing) or deterioration with increasing of operation and maintenance costs may be introduced. The improvements on operations (e.g., *learning curve*), introduction of multiple types of technology (e.g., energy production), the discrete enumeration of alternatives, and the long term uncertainty are being addressed through stochastic frameworks.

The Two-stage Stochastic Programming (2SSP) is widely used to address uncertainty. For instance, considering the *capacity expansion* problem: at the first stage, the capital and investment decisions must be taken, "here-and-now"; then, the project variables are obtained (project stage); in the second stage (recourse stage), the uncertainty is introduced through the set of foreseen scenarios and respective probability; this way, the control variables are calculated within a probabilistic character.

However, under uncertainty it is pertinent to promote robustness, either the *robustness on solution*—the solution remain "almost optimal" even when all scenarios are considered—and the *robustness in the model*—the optimal solution do not present high values to the excess/unused capacity or the unsatisfied demand. Usually a SLP objective function is modified in RO models by introducing penalization parameters on deviation, non-satisfied demand, capacity excess, or probabilistic restrictions are modified by enlarging/narrowing "soft" bounds.

Furthermore, using a theoretical approach and developing computational complexity studies of the PP problem, it is shown that most of the PP problems are NP-hard (Miranda 2007). Then, the necessity to develop heuristics may arise in case large instances are presenting. Nevertheless, it is suitable to solve instances of reasonable dimension using detailed knowledge about the models, by balancing their benefits and limitations. In the medium and large horizons, it is also pertinent to promote robustness in face of uncertainty. A generalization was then developed to properly address the "capacity production", to economically evaluate it, by considering flexible production schemes, and targeting robustness in 2SSP context. Logistics and financial subjects were not addressed in this generalization approach (e.g., inventory, distribution network, finance risk).

The purposes of the generalized model for processes planning thus are:

- To treat uncertainty and to promote robustness;
- To consider flexible processes, by integrating the production schemes formulation;
- To define the time implementation and the size of processes;
- To model economies of scale;
- To estimate the fluxes in the chemical processes, the raw materials' purchases, and the products sales.

The related nomenclature follows:

Index and sets

NC Number of components/materials *j*;
NM Number of markets *l* for purchase/sale materials;
NP Number of processes *i*;
NR Number of scenarios *r*;
NS Number of production schemes *s*;
NT Number of time periods *t*.

Parameters

$prob_r$ Probability associated with each scenario r;
λdsv, λzp Penalization parameters, respectively, for the solutions variability and the capacity slackness;
γ_{jlt}, Γ_{jlt} Unit prices for sale and purchase of the *j* materials, in market *l*, period *t*;
δ_{ist} Unit costs for the production processes *i*, at scheme *s*, period *t*;
α_{it}, β_{it} Variable and fixed costs for the capacity expansion of processes *i*, in each period *t*;
a_{jlt}, d_{jlt} Availabilities and demands of components and materials *j*, in market *l*, period *t*;

μ_{ijs} Characteristic constants that associate the inflows and outflows of the various components within each process i;

ρ_{is} Relative production rates at each process i, scheme s;

H_{it} Upper value for the availability of production time, for process i during period t;

$PS(i)$ Set of production schemes s allowed for flexible process i;

$QE_{it\tau}^{acum}$ Upper bound for the capacities to expand, concerning the aggregated "demand of capacities" from period t until period τ.

Variables

ξ_r Net Present Value (NPV) in scenario r;

$dsvn_r$ Negative deviation on the value of NPV in scenario r;

Sal_{jltr} Products sales, for material j, in market l, period t and scenario r;

Pur_{jltr} Materials purchases, for material j, in market l, period t and scenario r;

QE_{it} Capacity expansion of process i in period t;

Q_{it} Capacity of process i in period t;

y_{it} Binary decision related to the expansion of process, in period t;

Zp_{itr} Deviation by capacity slackness of process i, in period t at scenario r;

θ_{istr} Amount of principal component j' being processed at scheme s of process i, at period t and for each scenario r.

Then the model ***ROplan_flex*** (adapted from Miranda and Casquilho (2008) along with some examples) in the relations set (RO.1-a–RO.1-l) considers and simultaneously conjugates the significant points of prior PP models;

- A robust objective function (RO.1-a); and
- The flexibility schemes within the processes planning frame are addressed on restrictions set (RO.1-b–RO.1-l).

These relations are synoptically described:

(a) Objective Function, aiming to maximize the expected Net Present Value (NPV), considering penalizations on variability around the expected value, and penalizations on capacity slacks (non-statistical measure);
(b) Definition of NPV;
(c) Definition of the solution variability through the negative linear deviation;
(d) Definition of capacity slacks;
(e) Upper bound to the investment budget;
(f) Logic bounds onto the expansion of processes;
(g) Balance to the process capacity 's "production";
(h) Mass balance onto the components;
(i) Upper bounds to the materials purchases;
(j) Upper bounds to components sales;
(k) Non-negativity restrictions;
(l) Binary variables definition.

Model *ROplan_Flex:*

$$[\max]\Phi = \sum_{r=1}^{NR} prob_r \xi_r - \lambda dsv \sum_{r=1}^{NR} prob_r.dsvn_r - \lambda zp \sum_{i=1}^{NP}\sum_{t=1}^{NT}\sum_{r=1}^{NR} Zp_{itr} \quad \text{(RO.1-a)}$$

subject to

$$\begin{aligned}\xi_r = &\sum_{j=1}^{NC}\sum_{l=1}^{NM}\sum_{t=1}^{NT}\left(\gamma_{jlt} Sal_{jltr} - \Gamma_{jlt} Pur_{jltr}\right) \\ &- \sum_{i=1}^{NP}\sum_{s\in PS(i)}^{NS}\sum_{t=1}^{NT}(\delta_{ist}\rho_{is}\theta_{istr}) - \sum_{i=1}^{NP}\sum_{t=1}^{NT}(\alpha_{it} QE_{it} + \beta_{it} y_{it}), \quad \forall r\end{aligned} \quad \text{(RO.1-b)}$$

$$dsvn_r \geq \sum_{r'=1}^{NR}(prob_{r'}\xi_{r'}) - \xi_r, \quad \forall r \quad \text{(RO.1-c)}$$

$$\sum_{s\in PS(i)}^{NS} \theta_{istr} + Zp_{itr} = H_{it} Q_{it}, \quad \forall i,t,r \quad \text{(RO.1-d)}$$

$$\sum_{i=1}^{NP}(\alpha_{it} QE_{it} + \beta_{it} y_{it}) \leq CI(t), \quad \forall t \quad \text{(RO.1-e)}$$

$$QE_{it} \leq QE_{it}^{Upp} y_{it}, \quad \forall i,t \quad \text{(RO.1-f)}$$

$$Q_{it-1} + QE_{it} = Q_{it}, \quad \forall i,t \quad \text{(RO.1-g)}$$

$$\sum_{l=1}^{NM} Pur_{jltr} + \sum_{i=1}^{NP}\sum_{s\in PS(i)}^{NS} \mu_{ijs}\rho_{is}\theta_{istr} = \sum_{l=1}^{NM} Sal_{jltr}, \quad \forall j,t,r \quad \text{(RO.1-h)}$$

$$Pur_{jltr} \leq a_{jltr}^{Upp}, \quad \forall j,l,t,r \quad \text{(RO.1-i)}$$

$$Sal_{jltr} \leq d_{jltr}^{Upp}, \quad \forall j,l,t,r \quad \text{(RO.1-j)}$$

$$\xi_r,\ dsvn_r,\ Sal_{jltr},\ Pur_{jltr},\ \theta_{istr},\ Zp_{itr},\ QE_{it},\ Q_{it} \geq 0, \quad \forall i,j,l,s,t,r \quad \text{(RO.1-k)}$$

$$y_{it} \in \{0,1\}, \quad \forall i,t \quad \text{(RO.1-l)}$$

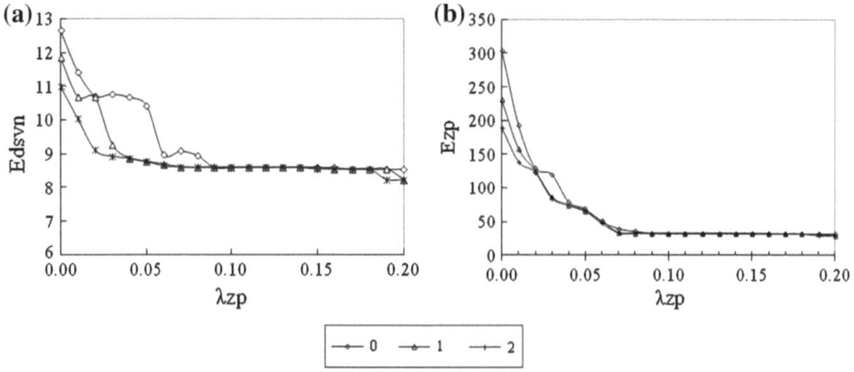

Fig. 1 Sensitivity of the objective function in face of the penalization parameters, showing solution robustness (**a**) and robustness in model (**b**)

Various examples were developed, and the sensitivity of the objective function considering the penalization parameters evolution is studied in the very first one. The graphs ***a*** and ***b*** in Fig. 1 consider:

(a) The variation of the expected value of the partial negative deviation, *Edsvn*, within the penalization parameters λzp and $\lambda dsvn$, showing the solution robustness; for all the analyzed values of $\lambda dsvn$ on the 0–2 range (only three lines are presented for easier observation), the expected value measuring the solution variability, *Edsvn*, reduces rapidly with the λzp evolution and a flat level is reached;
(b) The variation of the expected value of capacity slacks, *Ezp*, within the penalization parameters λzp and $\lambda dsvn$ showing robustness on model; analyzing again the 0–2 range for $\lambda dsvn$, the expected value measuring the capacity slackness, *Ezp*, also reduces rapidly with the λzp evolution onto a flat level.

In the second example, the issue of a reagent limitation is incorporated in the expansion of the existing dedicated process. Improvements of about 0.3 and 15 % due to the new two flexible processes are obtained, the superstructure and the average flows of the selected processes are presented in Fig. 2. Flexible processes are able to produce different components under different production schemes and different reagents, while only one configuration production-parameters-reagent is allowed on dedicated processes.

The superstructure represents a processes network with five continuous processes: the existing dedicated chemical process P3, produces high value component C7 using components C3 and C4, but component C3 has high price and short availability. The production of C3 is to be evaluated, and for that components C1 and C2 are to be purchased and transformed by chemical processes P1 and P2; these two processes are to be implemented if economically suitable. The flexible process P1 allows two production schemes, the first scheme presents similar parameters as

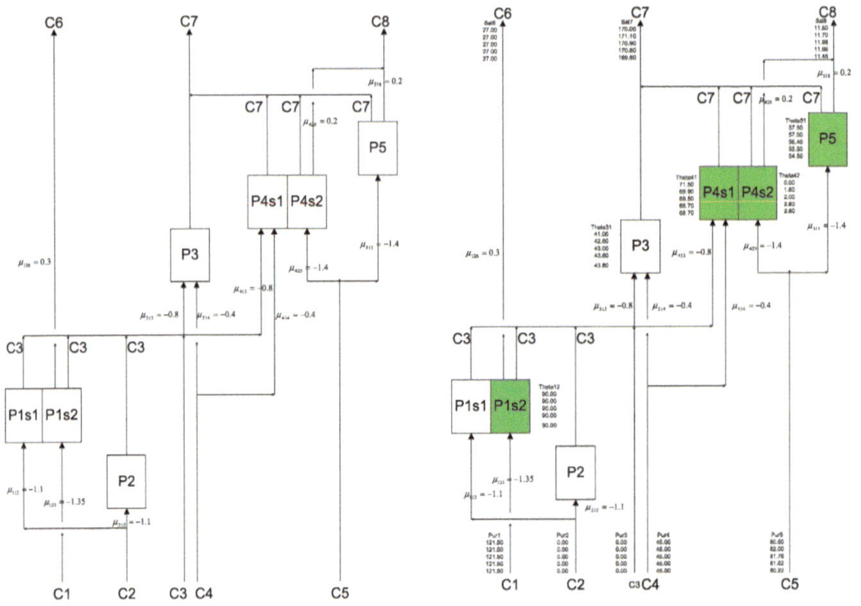

Fig. 2 Superstructure (*left*) and selection of flexible processes (*right*) for the numerical example on reagent limitation

dedicated process P2, but the second production scheme of P1 also allows the production of component C6. Component C6 presents high market value, so as components C7 and C8. In plus, the expansion of process P3 shall be evaluated too, and compared with the implementation of a flexible process P4 and/or the dedicated process P5: the parameters of the two production schemes on P4 are similar to the parameters of both dedicated processes P3 and P5. The uncertainty on components prices and demands along the time horizon is described by a discrete number of scenarios, and scenario probabilities are assigned.

A comparative study of the different production pathways can thus be developed, and the parameters that make economically preferable processes can be estimated. From the above, an alternative pathway for the production of component C7 is shown of interest, also including C8 production and using the replacement reagent C5. The self-production of limiting reagent C3 proved economically favorable too, preferably by the scheme that also produces C6 over the pathway which produces only C3. The selected processes are colored in the figure.

2.1 *Concluding Remarks on the Robust Process Planning*

The generalized process planning model, ***ROplan_flex***, integrated the risk policy of the decision maker. The model also integrates processes flexibility, the formulation of production schemes, and robustness without losing linearity. In addition, the expected values on deviation estimators, solution variability and capacity slacks fit the evolution the penalization parameters: consequently, the ***ROplan_flex*** model and the solutions are robust.

3 Robust Model for Design and Scheduling of Batch Processes

This section addresses the design of batch chemical processes and simultaneously considers the scheduling of operations. A model generalization is proposed, from a deterministic MILP model (Voudouris and Grossman 1992) to a stochastic 2SSP approach, this generalization being based on computational complexity studies (Miranda 2011a). The generalized model for design and scheduling of batch chemical process treats different time ranges, namely, the investment and scheduling horizons. Furthermore, the 2SSP framework allows the promotion of robustness solution, by penalizing the deviations; and the robustness in the model, with relaxation of the integrality constraints in the second phase variables.

In comparison with the evolution of PP models, a different way of evolution is observed for the design and scheduling problem (Miranda 2007). A realistic formulation was focused, with successive incorporation of freedom degrees, namely, extending the number of processes in each stage, the production and storage policies, and considering the economic charges on setups, operations, and inventories. Since the model becomes more and more difficult to solve, the solution procedures were adjusted in the early phases of modeling, and even network subjects were introduced to improve short term scheduling. Nevertheless, a minor impact of this case is noted, both in literature and industry applications.

In order to better select the equipment to purchase, the optimal production policy must also be found since it directly affects the equipment sizing. However, it involves the detailed resolution of scheduling subproblems where decomposition schemes are pertinent. These subproblems are focused in the second phase of 2SSP, where the control variables (recourse) occur. The integer and binary variables related to the scheduling and precedence constraints are disregarded as control variables, as they would make very hard the treatment of the recourse problem. Consequently, the second phase variables are assumed continuous (for example, the number of batches) and binary variables occur only in the first phase.

The study of existing models in the literature induces the enlargement of models and related applications (Miranda 2007), and this generalization of models simultaneously increases complexity and solution difficulties. A design and scheduling MILP model (Voudouris and Grossman 1992) that seems to have no improvements for more than a decade is addressed. Analytical results and computational complexity techniques were applied to the referred deterministic and single time MILP model, which is featuring multiple machines per stage, *zero wait* (ZW) and *single product campaign* (SPC) policy. That model was selected (Miranda 2007) because:

- For industrial applications with realistic product demands, multiple processes in parallel at each stage shall be considered, else numerical unfeasibility will occur;
- Due to modeling insufficiency, the option for the SPC policy arises from the difficulties to apply *multiple product campaign* (MPC) in a *multiple machine* environment;
- Assuming SPC, the investment cost will exceed in near 5 % the cost of the more efficient MPC policy; then, the SPC sizing is a priori overdesigned, and this will permit to introduce new products or to accommodate unforeseen growth on demands.

The generalized model ***RObatch_ms*** (adapted from Miranda and Casquilho (2011) along with some examples) includes the optimization of long term investment and also considers the short term scheduling of batch processes. Indeed, deterministic models do not conveniently address the risk of a wider planning horizon, and scheduling decision models often deal with certain data in a single time horizon. Thus, difficulty increases when the combinatorial scheduling problem is integrated with the uncertainty of the design problem. The following nomenclature is assumed:

Nomenclature

Index and sets

M	Number of stages i
NC	Number of components or products j
$NP(i)$	(Cardinal) number of processes $p(i)$ per stage
NR	Number of discrete scenarios r
$NS(i)$	(Cardinal) Number of discrete dimensions $s(i)$ in the process of stage i
NT	Number of time periods t;

Parameters

τ_{ij}	Processing times (h), for each product j in stage i
λdvt	Negative deviation on NPV penalization parameter
λqns	Non-satisfied demand penalization parameter
λslk	Capacity slack penalization parameter
c_{ips}	Equipment cost related to process $p(i)$ and size $s(i)$ selected in stage i
dv_{ij}	Discrete equipment volume in each stage
H	Time horizon
nc^{Upp}	Upper limit for disaggregated number of batches
$prob_r$	Probability of scenario r
$p(i)$	(Ordinal) Number of processes in stage i
Q_{itr}	Demand quantities (uncertain) for each product i

ret_{jtr} Unit (uncertain) values of return (net values) of the products j, in period t and scenario r
$s(i)$ (Ordinal) Number of process discrete dimensions in stage i
S_{ij} Dimension factor (L/kg), for each product j in stage i
V_{ij} Equipment volume (continuous value) in each stage;

Variables

ξ_r NPV value in scenario r
$dvtn_r$ Negative deviation on the value of NPV in scenario r
n_{jtr} Number of batches of product j, in period t and scenario r
nc_{ijsptr} Number of batches of product j, in period t and scenario r, disaggregated by process $p(i)$ and size $s(i)$ in each stage i
Qns_{jtr} Non-satisfied demand quantities of product j, in period t and scenario r
slk_{ijtr} Capacity slacks in each stage i, concerning totality of the *batches* of each product j, in period t and scenario r
$tcamp_{jtr}$ Campaign times (SPC) relative to each product j
W_{jtr} Global quantities produced of product j, in period t and scenario r
y_{isp} Binary decision related to process $p(i)$ and size $s(i)$ selected in stage i.

Then the model ***RObatch_ms*** in the relations set (RO.2-a–RO.2-m) considers a robust objective function (***a***) and the design and scheduling of batch plants with multiple machines and SPC policy are addressed on restrictions set (RO.2-b–RO.2-m) as follows:

(a) Objective Function, aiming to maximize the expected Net Present Value (NPV), considering penalizations of variability and capacity slackness;
(b) Definition of NPV—Each probabilistic part ξ corresponds to the NPV obtained at each discrete scenario r, and this part is obtained as the present amount of sales return less the investment costs;
(c) Definition of the solution variability through the negative linear deviation;
(d–e) The non-satisfied demand of each product, *Qns*, is related with the constraint slack, being a non-negative variable by definition;
(f) The global excess (*slk*) on the implemented production capacities results directly from the slacks of the constraints on the global quantities produced for each product;
(g–i) The disaggregated number of batches, *nc*, corresponds to the product-aggregation's variables (*n*. *y*), and the three logical sets of constraints consider: (g) upper bounds; (h) only one value is selected; and (i) the specification of its selected value;
(j–k) The campaign times, *tcamp*, must be determined to satisfy the time horizon, *H*, and are related to the number of batches, *nc*;
(l) Non-negativity restrictions;
(m) Binary variables definition.

Model *RObatch_ms*:

$$[\max]\Phi = \sum_{r=1}^{NR} prob_r \xi_r - \lambda dvt \sum_{r=1}^{NR} prob_r.dvtn_r - \lambda qns \sum_{r=1}^{NR} \frac{prob_r}{NC.NT}\left(\sum_{j=1}^{NC}\sum_{t=1}^{NT} Qns_{jtr}\right)$$
$$- \lambda slk \sum_{r=1}^{NR} \frac{prob_r}{M.NC.NT}\left(\sum_{i=1}^{M}\sum_{j=1}^{NC}\sum_{t=1}^{NT} slk_{ijtr}\right) \quad \text{(RO.2-a)}$$

subject to,

$$\xi_r = \sum_{j=1}^{NC}\sum_{t=1}^{NT} ret_{jtr} W_{jtr} - \sum_{i=1}^{M}\sum_{s=1}^{NS(i)}\sum_{p=1}^{NP(i)} c_{isp} y_{isp}, \quad \forall r \quad \text{(RO.2-b)}$$

$$dvtn_r \geq \sum_{r'=1}^{NR} (prob_{r'} \xi_{r'}) - \xi_r \geq 0, \quad \forall r \quad \text{(RO.2-c)}$$

$$W_{jtr} + Qns_{jtr} = Q_{jtr}, \quad \forall j, t, r \quad \text{(RO.2-d)}$$

$$Qns_{jtr} \geq 0, \quad \forall j, t, r \quad \text{(RO.2-e)}$$

$$S_{ij} W_{jtr} + slk_{ijtr} = \sum_{s=1}^{NS(i)}\sum_{p=1}^{NP(i)} dv_{is} nc_{ijsptr}, \quad \forall i, j, t, r \quad \text{(RO.2-f)}$$

$$nc_{ijsptr} - nc_{ijsp}^{Upp} y_{isp} \leq 0, \ \forall\, i, j, s, p, t, r \quad \text{(RO.2-g)}$$

$$\sum_{s=1}^{NS(i)}\sum_{p=1}^{NP(i)} y_{isp} = 1, \quad \forall i \quad \text{(RO.2-h)}$$

$$\sum_{s=1}^{NS(i)}\sum_{p=1}^{NP(i)} nc_{ijsptr} - n_{jtr} = 0, \quad \forall i, j, t, r \quad \text{(RO.2-i)}$$

$$\sum_{s=1}^{NS(i)}\sum_{p=1}^{NP(i)} \left(\frac{\tau_{ij}}{p(i)} nc_{ijsptr}\right) - tcamp_{jtr} \leq 0, \quad \forall i, j, t, r \quad \text{(RO.2-j)}$$

$$\sum_{j=1}^{NC} tcamp_{jtr} \leq H, \quad \forall t, r \quad \text{(RO.2-k)}$$

Table 1 Numbers of parameters, variables and constraints corresponding to examples solved assuming: NC = 4; M = 3; NS = 5; NP = 3; NT = 1

Numerical examples	Parameters *NR*	Continuous variables	Constraints
EX1	1	209	226
EX2	3	603	672
EX3	7	1391	1564
EX4	15	2967	3348
EX5	30	5922	6693
EX6	100	19712	22303

$$\xi_r, dvtn_r, slk_{ijtr}, nc_{ijsptr}, n_{jtr}, Qns_{jtr}, tcamp_{jtr}, W_{jtr} \geq 0, \quad \forall i,j,s,p,t,r \qquad \text{(RO.2-l)}$$

$$y_{isp} \in \{0; 1\}, \quad \forall i,s,p \qquad \text{(RO.2-m)}$$

The application of the generalized model ***RObatch_ms*** for the robustness promotion is illustrated through various numerical examples, as indicated in Table 1.

For that, numerical instances of uncertain demands on a unique time period ("static") are addressed, and the generalization of the deterministic problem to the minimization of investment costs in a stochastic and robust formulation is considered. Through the usual reasoning of polynomial reduction of problem instances, the following is assumed: (i) only one time period; and (ii) zero value of return in products.

Considering one single time period, the unsuitability of NPV maximization must be noticed: NPV is usually addressed in a multiperiod horizon (dynamic problem) due to high investment costs that do not allow payback at one single time period. Furthermore, supposing $ret_{jtr} = 0$, then the NPV variables ξ are representing only the investment costs because there are no cash flows returning:

$$\xi = \xi_r = -\sum_{i=1}^{M} \sum_{s=1}^{NS(i)} \sum_{p=1}^{NP(i)} c_{isp} y_{isp}, \quad \forall r \qquad \text{(RO.3)}$$

There is no variability at this instance: NPV variables ξ are scenario independent and a null deviation, $dvtn_r = 0$, is observed in all scenarios. And aiming to satisfy the uncertain product demands, the penalization of capacity slacks is not being considered ($\lambda slk = 0$). The objective function in equation RO.2-a is thus reduced to the robust minimization of investment costs, assuming only the penalization of non-satisfied demand:

$$\begin{aligned}[\max]\Phi &= \xi.\sum_{r=1}^{NR} prob_r - \lambda qns \sum_{r=1}^{NR} \frac{prob_r}{NC.NT}\left(\sum_{j=1}^{NC}\sum_{t=1}^{NT} Qns_{jtr}\right) \\ &= -\sum_{i=1}^{M}\sum_{s=1}^{NS(i)}\sum_{p=1}^{NP(i)} c_{isp}y_{isp} - \lambda qns \sum_{r=1}^{NR} \frac{prob_r}{NC.NT}\left(\sum_{j=1}^{NC}\sum_{t=1}^{NT} Qns_{jtr}\right)\end{aligned} \quad \text{(RO.4)}$$

or,

$$[\min]\Psi = \sum_{j=1}^{M}\sum_{s=1}^{NS(i)}\sum_{p=1}^{NP(i)} c_{isp}y_{isp} + \lambda qns \sum_{r=1}^{NR} \frac{prob_r}{NC.NT}\left(\sum_{j=1}^{NC}\sum_{t=1}^{NT} Qns_{jtr}\right) \quad \text{(RO.5)}$$

The model defined in the relations set (RO.2-a–RO.2-m) is being restricted, and the following can be neglected: the time period index, *t*, because only one time period is considered; the constraint sets concerning the definition of the probabilistic variables, ξ, which will have a constant and scenario-independent value; and, the same for the deviation definitions, *dvtn*, which consequently will be null and useless. The various examples (EX1 to EX6) are described in Table 1.

The effect related to the utilization of distinct numbers of discrete scenarios is analyzed in the generalized stochastic model, which conceptually reduces to the deterministic model when considering one single scenario. Although the number of binary variables is kept constant and equal to 45 for all the six examples, both the number of continuous variables and the number of constraints vary linearly with the number of scenarios, *NR*.

(a) Variation of robust costs (10^5 €) versus the penalty values on non-satisfied demand, λqns, for various numerical examples.
(b) Variation of the expected value for non-satisfied demand, *Ensd* (10^3 kg), with the evolution of λqns, for various numerical examples.

Graphical representations are shown for the variation of different estimators (robust cost, ψ, expected value of the non-satisfied demand, *Ensd*, expected value of the capacity slacks, *Eslk*, non-robust cost, *Ecsi*), with the conjugated increase of the number of scenarios, *NR*, and the penalization for non-satisfied demand, λqns. Due to the near coincidence of the different lines represented, from *NR* = 1 (EX1) to *NR* = 100 (EX6), only three lines are shown in these graphs.

Considering the graphs at Fig. 3, two key subjects are observed:

(i) the behavior of the numerical instances is similar even when different number of scenarios is considered, *NR* from 1 to 100; and
(ii) the model robustness, with adequate sensitivity of technical estimators to the evolution of the non-satisfied demand penalization parameter, λqns; for all the scenarios number on the 1–100 range, the expected values associated with robust cost, ψ, and the non-satisfied demand, *Ensd*, alter rapidly with the λqns evolution and a flat level is reached in both cases.

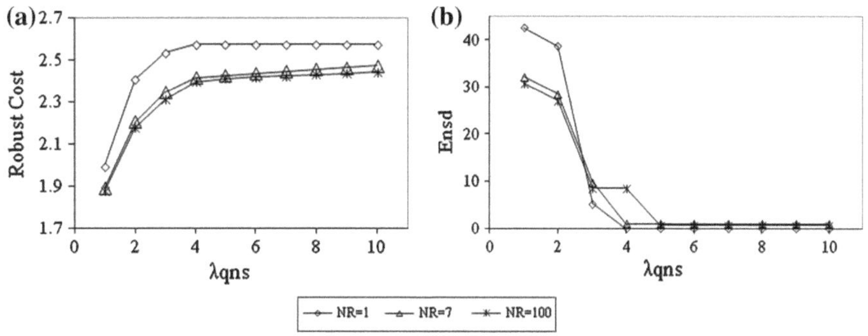

Fig. 3 Sensitivity of the objective function in face of the penalization parameters, showing solution robustness (**a**) and robustness in model (**b**)

Table 2 Significant values of the robust optimization considering distinct instances of numerical example EX3 (for *NR* = 7)

λqns	Costs	Ord(s)	Sum(dv)	%Eslk	%Ensd
0	124,596.08	1/1/1	3451.47	21.0	54.1
1	156,737.60	1/3/3	6087.68	11.7	21.0
2	163,661.91	1/3/4	6659.83	11.5	18.6
3	205,467.00	2/5/5	9366.59	11.1	6.3
4	237,689.36	3/5/5	10,298.62	13.6	0.6
5	237,689.36	3/5/5	10,298.62	12.0	0.6
	(…)		(…)		(…)
10	237,689.36	3/5/5	10,298.62	13.2	0.6
			(…)		(…)
20	237,689.36	3/5/5	10,298.62	12.7	0.6
21	257,332.14	4/5/5	10,929.24	13.3	0.0
	(…)		(…)		(…)
40	257,332.14	4/5/5	10,929.24	12.8	0.0

In Table 2, significant values for instances of EX3 are shown, in which instances are considered with seven scenarios, but whose variation with the growth of the penalization on the non-satisfied demand, λqns, is similar to all the other examples with different numbers of scenarios. These significant values are the values associated to alterations in the optimum configuration. Table 2 relates the penalty parameters λqns and the (non-robust) costs, showing:

- the order of the discrete dimension (size, *s*) selected in each stage, *Ord*(*s*); for example, "4/7/2", indicates that the fourth dimension was chosen at the first stage, the seventh dimension at the second stage, and the second dimension at the third stage; these values come directly from the binary solution;

- the sum of the discrete dimensions or equipment volumes selected, *Sum*(*dv*); this measure is important for slackness analysis, due to the interest of appreciating *Eslk* in relative terms at each example;
- the percentage of non-satisfied demand expected value, *%Ensd*; for each example, the reference value (100) is the expected value of uncertain demand;
- the percentage of capacity slacks expected value, *%Eslk*; for each example, the reference value for the calculation of this estimator is the discrete volumes sum, *Sum*(*dv*).

In Table 2, a progressive increase on costs is observed, due to the corresponding increase of the orders of dimension, *Ord*(*s*), and the increasing sum of volumes, *Sum*(*dv*). But in percentage of the expected value of the capacity slacks, *%Eslk*, it is verified that this one is kept in a strict range of values, between 11 % and 13 %. In absolute terms, the expected value for the slack, *Eslk*, is increasing for the first values of the penalty, λqns, but then attains a stable value. This permits to confirm the inherence of a residual value for the capacity slacks in this type of problem, where multiple products are processed in the same equipment. Focusing the effect of non-satisfied demand penalization, λqns, on the related percent expected value, *%Ensd*:

- if the penalization parameter is large, all the demand will be satisfied, *Ensd* and *%Ensd* tend to be null; the selected dimensions and the related costs are also large, but notice that one single scenario with tiny probability can drive such a large sizing;
- instead of full demand satisfaction, if the requirement of non-satisfied demand is relaxed to 1 %, the investment costs are reduced in about 8 %; and assuming more flexibility, if the non-satisfied demand is allowed to reach up to 6 %, the cost reduction is about 20 %;
- this kind of reasoning is realistic, since the model and the numerical examples are based on SPC policy; and if SPC is considered instead of MPC in the design and scheduling models studied, a relative overdesign of about 5 % in investment costs is foreseen (Miranda 2007).

3.1 Concluding Remarks on Robust Design and Scheduling

The MILP model featuring SPC and multiple machines (Voudouris and Grossmann 1992) was found to be the most promising from a computational standpoint (Miranda 2011b), and it was generalized toward a stochastic model ***RObatch_ms***. Its results point to a significant reduction (8–20 %) on the investment costs in comparison to the deterministic non-relaxed case. If the MPC policy is adopted or if a slight relaxation is made to the impositions on the uncertain demands (respectively, of 1–6 %), the demand relaxation does not cause real losses: model ***RObatch_ms*** assumes the lower efficient SPC policy and an overdesign in about

5 % on investment cost is thus expected (Miranda 2007). And remark also the ***RObatch_ms*** model robustness, with estimators responding adequately to the variation of the penalty parameter for non-satisfied demand, λqns.

4 Conclusions

The study of optimization cases and models from the literature allowed a detailed overview, and permitted to conjugate realistic subjects both in formulation and solution procedures. Developing theoretical studies, the various models at hand are detailed and insight is gained, their benefits and limitations are balanced, and robust generalization is developed. In addition, the studies on computational complexity along with the computational implementation fostered the construction of heuristics, such as local search procedures.

Based on the generalization approach described in prior paragraph, two problems are addressed:

- The PP problem—beyond the uncertainty treatment it is also considered the formulation of flexible production schemes, the robustness on solution and on model, and the processes parameters were economically evaluated.
- The *Batch* problem—the treatment of uncertainty also considered the problems' specificities; the short term scheduling and the multi-period horizon were simultaneously addressed, the deterministic approach from the literature is generalized onto a stochastic one, and economic parameters of interest were evaluated.

Further developments include modeling issues and solution methods, while the development of Decision Support Systems will foster the application to industrial cases.

Acknowledgments Authors thank the College of Technology and Management at the Portalegre Polytechnics Institute (ESTG/IPP) and Instituto Superior Técnico (IST). These works were partially developed at the Centre for Chemical Processes (CPQ/IST) with the support of FCT project PEst-OE/EQB/UI0088, and other developments at CERENA/IST with the support of FCT project UID/ECI/04028/2013. We also thank reviewers' comments that helped to improve the text.

Appendix 1: Technical and Economic Estimators

The non-robust NPV expected value:

$$Ecsi = \sum_{r=1}^{NR} prob_r \xi_r \tag{A.1}$$

The negative deviation expected value:

$$Edvt = \sum_{r=1}^{NR} prob_r.dvtn_r \tag{A.2}$$

The non-satisfied demand expected value:

$$Ensd = \sum_{r=1}^{NR} \frac{prob_r}{NC.NT}\left(\sum_{j=1}^{NC}\sum_{t=1}^{NT} Qns_{jtr}\right) \tag{A.3}$$

The capacity slack expected value:

$$Eslk = \sum_{r=1}^{NR} \frac{prob_r}{M.NC.NT}\sum_{j=1}^{NC}\sum_{t=1}^{NT}\left\{\sum_{i=1}^{M}\sum_{s=1}^{NS}\sum_{p=1}^{NP} p(i).y_{isp}.\left(dv_{js} - S_{ij}.\frac{W_{jtr}}{n_{jtr}}\right)\right\} \tag{A.4}$$

The percentage non-satisfied demand expected value:

$$\%Ensd = \frac{Ensd}{Qmed}.100, \quad \text{with} \quad Qmed = \sum_{r=1}^{NR} \frac{prob_r}{NC.NT}\left(\sum_{j=1}^{NC}\sum_{t=1}^{NT} Q_{jtr}\right) \tag{A.5}$$

The percentage capacity slack expected value:

$$\%Eslk = \frac{Eslk}{Vtotal}.100, \quad \text{with} \quad Vtotal = \sum_{i=1}^{M}\sum_{s=1}^{NS}\sum_{p=1}^{NP}\left(y_{isp}.dv_{is}\right) \tag{A.6}$$

References

Bok, J. K., Lee, H., & Park, S. (1998). Robust investment model for long-range capacity expansion of chemical processing networks under uncertain demand forecast scenarios. *Computers & Chemical Engineering, 22*, 1037–1049.

Garey, M. R., Johnson, D. S., & Sethi, R. (1976). The complexity of flowshop and jobshop scheduling. *Mathematics of Operations Research, 2*, 117–129.

Miranda, J. L. (2007). Optimização em Sistemas de Processos Químicos: Generalização de Modelos com Planeamento e Sequenciamento. PhD. Thesis, Instituto Superior Técnico, Universidade Técnica de Lisboa.

Miranda, J. L. (2011a). Computational complexity studies in the design and scheduling of batch processes. In: Book of Abstracts of IO2011—15th Congress of APDIO, University of Coimbra, Coimbra, Portugal, 18–20 April 2011.

Miranda, J. L. (2011b). The design and scheduling of chemical batch processes: building heuristics and probabilistic analysis. *Theory and Applications of Mathematics & Computer Science, 1*(1), 45–62.

Miranda, J. L., & Casquilho, M. (2008). Abordagem generalizada ao planeamento de processos de produção: flexibilidade e robustez. *Engenharia Química, 8*, 26–33.

Miranda, J. L., & Casquilho, M. (2011). Design and scheduling of chemical batch processes: generalizing a deterministic to a stochastic model. *Theory and Applications of Mathematics & Computer Science, 1*(2), 71–88.

Miranda, J. L., & Nagy, M. (2011). A case of cooperation in the European Operations Research education. *European Journal of Engineering Education, 36*(6), 571–583.

Norton, L. C., & Grossmann, I. E. (1994). Strategic planning for complete process flexibility. *Industrial & Engineering Chemistry Research, 33*, 69–76.

Sahinidis, N. V., Grossmann, I. E., Fornari, R. E., & Chathrathi, M. (1989). Optimization model for long range planning in the chemical industry. *Computers & Chemical Engineering, 13*, 1049–1063.

Voudouris, V. T., & Grossmann, I. E. (1992). Mixed-Integer linear programming reformulations for batch processes design with discrete equipment sizes. *Industrial and Engineering Chemistry Research, 31*, 1315–1325.

Multiobjective Interval Transportation Problems: A Short Review

Carla Oliveira Henriques and Dulce Coelho

Abstract The conventional transportation problem usually involves the transportation of goods from several supply points to different demand points and considers the minimization of the total transportation costs. The transportation problem is a special case of linear programming models, following a particular mathematical structure, which has a wide range of potential practical applications, namely in logistic systems, manpower planning, personnel allocation, inventory control, production planning and location of new facilities. However, in reality, the transportation problem usually involves multiple, conflicting, and incommensurable objective functions, being called the multiobjective transportation problem. Several methods have been developed for solving this sort of problems with the assumption of precise information regarding sources, destinations and crisp coefficients for the objective function coefficients. Nevertheless, when dealing with real-life transportation problems, these circumstances may not be verified, since the transportation costs may vary as well as supply and demand requirements. Therefore, different approaches for dealing with inexact coefficients in transportation problems have been proposed in scientific literature, namely with the help of fuzzy and interval programming techniques. This paper is aimed at providing a short critical review of some interval programming techniques for solving this particular type of problems.

C. Oliveira Henriques (✉)
Polytechnic Institute of Coimbra - ISCAC Business School, Quinta Agrícola, Bencanta, 3040-316 Coimbra, Portugal
e-mail: chenriques@iscac.pt
URL: http://www.iscac.pt

D. Coelho
Polytechnic Institute of Coimbra - ISEC, Rua Pedro Nunes, Quinta da Nora, 3030-199 Coimbra, Portugal
e-mail: dcoelho@isec.pt
URL: http://www.isec.pt

C. Oliveira Henriques · D. Coelho
INESC Coimbra, Coimbra, Portugal

A.P. Barbosa-Póvoa et al. (eds.), *Optimization and Decision Support Systems for Supply Chains*, Lecture Notes in Logistics,
DOI 10.1007/978-3-319-42421-7_7

Keywords Multiobjective transportation problems · Multiobjective interval programming · Interval order relations

1 Introduction

The transportation problem (TP) might be seen as a particular case of linear programming (LP) models which is generally used to determine the optimal solution for the distribution of certain goods, from different supply points (sources—e.g. production facilities, warehouses) to different demand points (destinations—e.g. warehouses, sales, outlet), considering that there is a certain distance between these points. The objective function in the TP usually represents the total transportation costs, while the constraints are defined by the supply capacity and demand requirements of certain sources or destinations, respectively. However, in real-life situations the TP usually encompasses multiple, conflicting and incommensurate objective functions (e.g. transportation cost, average delivery time, number of goods transported, unfulfilled demand). This type of problem is also known as the multiobjective transportation problem (MOTP). The conventional TP and MOTP follow a particular mathematical structure assuming that the coefficients of the objective functions and the supply and demand values are stated in a precise way, with crisp values. Nevertheless, such assumptions are rarely satisfied (e.g. the unit transportation costs may vary and the supply and demand may also change). Therefore, TP and MOTP for decision support must take explicitly into account the treatment of the inherent uncertainty associated with the model coefficients. Interval programming is one of the approaches to tackle uncertainty in mathematical programming models, which holds some interesting characteristics because it does not require the specification or the assumption of probabilistic distributions (as in stochastic programming) or possibilistic distributions (as in fuzzy programming) or a max-min formulation (as in robust optimization). The use of interval programming techniques is possible providing that information about the range of variation of some (or all) of the parameters is available (Oliveira and Antunes 2007).

This paper is aimed at providing an overview of the different approaches reported in scientific literature regarding interval programming techniques for tackling the uncertainty in TP and MOTP. This paper is structured as follows: Sect. 2 provides the underpinning assumptions of MOTP with interval coefficients (MOITP); Sect. 3 discusses the main approaches found in scientific literature for obtaining solutions to MOITP. Finally, Sect. 4 concludes highlighting the main advantages and drawbacks of the several approaches also suggesting possible future work development in this field of research.

2 Theoretical Underpinning of MOITP

Conventional multiobjective linear programming (MOLP)/LP models usually address practical problems in which all coefficients and parameters are a priori given. However, due to the inexactness and uncertainty aspects of these problems should be explicitly taken into consideration. Uncertainty handling can be dealt with in various ways, namely by means of stochastic, fuzzy and interval programming techniques. In the stochastic approach the coefficients are treated as random variables with known probability distributions. In the fuzzy approach, the constraints and objective functions are regarded as fuzzy sets with known membership functions. However, it is not always easy for the decision-maker (DM) to specify these probability distributions and membership functions. In the interval approach it is considered that the uncertain values are perturbed simultaneously and independently within known fixed bounds, being therefore intuitively preferred by the DM in practice.

2.1 Interval Numbers and Interval Order Relations

Consider that the value x (a real number) is uncertain, knowing that x lies between two real numbers a^L and a^U forming an interval, where $a^L < a^U$. All numbers within this interval have the same importance. An interval number A is defined as the set of real numbers x such that $a^L \leq x \leq a^U$, i.e. $x \in [a^L, a^U]$, $a^L, a^U \in \Re$ or,

$$A = \left[a^L, a^U\right] = \left\{x : a^L \leq x \leq a^U, \in \Re\right\}. \tag{1}$$

The width and midpoint of the interval number $A = [a^L, a^U]$ are $w[A] = (a^U - a^L)$ and $m[A] = \frac{1}{2}(a^U + a^L)$, respectively. There are basically two types of order relations between intervals: one based on the extension of the concept "<" (less than) for real numbers and another based on the extension of the concept of set inclusion. Consider the intervals $A = [a^L, a^U]$ and $B = [b^L, b^U]$ given in the set $I(\Re)$ i.e. the set of real interval numbers. Thus, it is considered that A (<) B if and only if $a^U < b^L$. Moreover, $A \subseteq B$ if and only if $a^L \geq b^L$ and $a^U \leq b^U$. Nonetheless, these interval order relations do not allow comparing overlapped intervals. On the other hand, the extension of set inclusion does not allow ordering intervals in terms of their importance. In this context, several approaches have been suggested which allow comparing two interval numbers. Ishibuchi and Tanaka developed an approach which allows comparing two interval numbers (Ishibuchi and Tanaka 1990). For instance, in a problem where the objective function is minimized, A is better than B, i.e. $A \leq_{LU} B$ if and only if $a^L \leq b^L$ and $a^U \leq b^U$.

The interval order relation "$\leq_{LU}$" has the following properties:

(a) If $A \leq_{LU} B$, then $m[A] \leq m[B]$.
(b) If $a^L = a^U$ and $b^L = b^U$, then "$\leq_{LU}$" corresponds to the inequality "$\leq$" used in the set of real numbers.

However, there are interval numbers that cannot be compared through the interval order relation "$\leq_{LU}$". For example, if $A = [85, 95]$ and $B = [80, 100]$, the order relations $A \leq_{LU} B$ and $B \leq_{LU} A$ are not verified. Therefore, Ishibuchi and Tanaka suggested another interval order relation based on the midpoint and width of the interval numbers, called "$\leq_{MW}$" (Ishibuchi and Tanaka 1990). According to this interval order relation, A is better than B, i.e. $A \leq_{MW} B$ if and only if $m[A] \leq m[B]$ and $w[A] \geq w[B]$.

The interval order relation "$\leq_{MW}$" has the following properties:

(a) If $A \leq_{MW} B$ then $a^L \leq b^L$.
(b) If $w[A] = w[B] = 0$, then "$\leq_{MW}$" corresponds to the inequality "$\leq$" used in the set of real numbers.

Ishibuchi and Tanaka have also defined another interval order relation, "$\leq_{LM}$", which allows comparing the previous two (Ishibuchi and Tanaka 1990). According to this interval order relation, A is better than B, i.e. $A \leq_{LM} B$ if and only if $a^L \leq b^L$ and $m[A] \leq m[B]$. On the other hand, $A \leq_{LM} B$ if and only if either $A \leq_{LU} B$ or $A \leq_{MW} B$. Chanas and Kuchta (1996) proposed a generalization of the interval order relations suggested by Ishibuchi and Tanaka (1990) introducing the concept of cutting level—φ_0, φ_1—of an interval. Consider the interval order relations "$\leq_{LU}$", "$\leq_{MW}$", "$\leq_{LM}$" defined as interval order relations of the type 1, 2 and 3, respectively. Let $A = [a^L, a^U]$ and φ_0 and φ_1 any real crisp numbers such that $0 \leq \varphi_0 \leq \varphi_1 \leq 1$. The cutting level—$\varphi_0$, φ_1—of interval A is:

$$A/\left[\varphi_0, \quad \varphi_1\right] = \left[a^L + \varphi_0\left(a^U - a^L\right), a^L + \varphi\left(a^U - a^L\right)\right]. \tag{2}$$

Let $A = [a^L, a^U]$ and $B = [b^L, b^U]$ be two interval numbers, φ_0 and φ_1 any real crisp numbers such that $0 \leq \varphi_0 \leq \varphi_1 \leq 1$ and i any interval order relation of the set $\{1, 2, 3\}$. The interval order relations $\leq_{i/\left[\varphi_0, \quad \varphi_1\right]}$ and $<_{i/\left[\varphi_0, \quad \varphi_1\right]}$ are defined in the following way:

$$A \leq_{i/\left[\varphi_0, \quad \varphi_1\right]} B \quad \text{if and only if} \quad A/\left[\varphi_0, \quad \varphi_1\right] \leq_i B/\left[\varphi_0 \quad \varphi_1\right], \tag{3}$$

$$A <_{i/\left[\varphi_0 \quad \varphi_1\right]} B \quad \text{if and only if} \quad A/\left[\varphi_0, \quad \varphi_1\right] <_i B/\left[\varphi_0 \quad \varphi_1\right] \tag{4}$$

where $A/\left[\varphi_0, \quad \varphi_1\right]$ and $B/\left[\varphi_0, \quad \varphi_1\right]$ correspond to the cutting levels—φ_0, φ_1—of interval numbers A and B, respectively.

Then, $A \leq_{i/\left[\varphi_0, \quad \varphi_1\right]} B$ if and only if $A \leq_i B, i = 1, 2, 3$ and $A \leq_{1/\left[\varphi_0, \quad \frac{\varphi_1}{2}\right]} B$ if and only if $A \leq_{3/\left[\varphi_0, \quad \varphi_1\right]} B$. Nevertheless, there are interval numbers that cannot be compared through the interval order relations defined by Ishibuchi and Tanaka (1990) and Chanas and Kuchta (1996). For example, let us consider $A = [1850, 2215]$ and $B = [1695, 2515]$. Therefore, $m[A] = 2032.5$, $m[B] = 2105$, $w[A] = 365$ and $w[B] = 820$. Therefore it might be concluded that interval A allows obtaining a lower value but less uncertain, whereas interval B allows obtaining a higher value but more uncertain. Thus, it is not possible to know with any of the interval order relations previously established what is the better interval value. In order to overcome this drawback Sengupta and Pal suggested an index which allows comparing any type of interval numbers, taking into account the satisfaction levels of the DM (Sengupta and Pal 2000). Consider the following expanded order relation $\prec$, between the interval numbers A and B. Sengupta and Pal (2000) defined an acceptability function $\mathcal{A} : I(\Re) \times I(\Re) \rightarrow [0, \infty]$, such that $\mathcal{A}(A \prec B)$ or $\mathcal{A} \prec (A, B)$, or even,

$$\mathcal{A} \prec = \frac{(\mathrm{m[B]} - \mathrm{m[A]})}{\left(\frac{\mathrm{w[B]}}{2} + \frac{\mathrm{w[A]}}{2}\right)}, \tag{5}$$

where $\left(\frac{w[B]}{2} + \frac{w[A]}{2}\right) \neq 0$. The index $\mathcal{A} \prec$ can be interpreted has the degree of acceptability of A being inferior to interval B, i.e. $\mathcal{A}(A \prec B)$ or $\mathcal{A} \prec (A, B)$.

The degree of acceptability of $A \prec B$ can be classified by comparing the mid-point and width of interval numbers A and B in the following way:

$$\mathcal{A}(A \prec B) = \begin{cases} 0 & \text{if} \quad \mathrm{m[A]} = \mathrm{m[B]}, \\ ]0,1[& \text{if} \quad \mathrm{m[A]} < \mathrm{m[B]} \quad \text{and} \quad \mathrm{a^U} > \mathrm{b^L}, \\ [1,\infty[& \text{if} \quad \mathrm{m[A]} < \mathrm{m[B]} \quad \text{and} \quad \mathrm{a^U} \leq \mathrm{b^L}. \end{cases} \tag{6}$$

If $\mathcal{A}(A \prec B) = 0$ then the premise "A inferior to B" is not accepted; if $0 < \mathcal{A}(A \prec B) < 1$ then the premise "A inferior to B" is accepted with different degrees of satisfaction; if $\mathcal{A}(A \prec B) \geq 1$ then the premise "A inferior to B" is accepted as true. In a minimization problem, if $\mathcal{A}(A \prec B) > 0$, then it should be concluded that A is better than B. The index $\mathcal{A} \prec$ satisfies the DM for any possible judgment regarding the comparison of any pair of interval numbers A and B, since it is always possible to obtain at least one of the following situations: $\mathcal{A}(A \prec B) > 0$, $\mathcal{A}(B \prec A) > 0$ or $\mathcal{A}(A \prec B) = \mathcal{A}(B \prec A) = 0$.

2.2 MOLP Models with Interval Coefficients and Parameters

Consider, without loss of generality, the following MOLP model with interval coefficients and parameters and the interval arithmetic operations (Oliveira and Antunes 2007; Moore 1966):

$$\begin{aligned} &\text{Maximize } \mathrm{Z_k}(\mathbf{x}) = \sum_{\mathrm{j=1}}^{\mathrm{n}} \left[\mathrm{c}^{L}_{\mathrm{kj}}, \ \mathrm{c}^{\mathrm{U}}_{\mathrm{kj}}\right] \mathrm{x_j} \quad (\mathrm{k} = 1, \ldots, \mathrm{p}), \\ &\text{subject to} \sum_{\mathrm{j=1}}^{\mathrm{n}} \left[\mathrm{a}^{\mathrm{L}}_{\mathrm{ij}}, \ \mathrm{a}^{\mathrm{U}}_{\mathrm{ij}}\right] \mathrm{x_j} \leq \left[\mathrm{b}^{\mathrm{L}}_{\mathrm{i}}, \ \mathrm{b}^{\mathrm{U}}_{\mathrm{i}}\right] \quad (\mathrm{i} = 1, \ldots, \mathrm{m}), \\ &\mathrm{x_j} \geq 0 \quad (\mathrm{j} = 1, \ldots, \mathrm{n}), \end{aligned} \tag{7}$$

where $\left[c_{kj}^{L},\ c_{kj}^{U}\right]$, $\left[a_{ij}^{L},\ a_{ij}^{U}\right]$ and $\left[b_{i}^{L},\ b_{i}^{U}\right]$, k = 1, …, p, j = 1, …, n, i = 1, …, m, correspond to closed intervals.

The interval programming approach has been used to tackle specific issues in MOLP. Some algorithms only tackle the uncertainty in the objective functions, others deal both with the uncertainty in the objective functions and in the right hand side (RHS) of the constraints and others handle with the uncertainty in all the coefficients of the model (Oliveira and Antunes 2007). According to Inuiguchi and Kume and Inuiguchi and Sakawa there are two different approaches to deal with an interval objective function: the satisficing approach and the optimizing approach (Inuiguchi and Kume 1994; Inuiguchi and Sakawa 1995). According to the satisficing approach each interval objective function is transformed into one or several objective functions (the lower bound, the upper bound and the midpoint of the intervals are usually used) in order to obtain a compromise solution. However, the use of this approach may lead to a compromise solution that might not be the most suitable one, if the gradients of the chosen objective functions are highly correlated (Antunes and Clímaco 2000). The optimizing approach extends the concept of efficiency used in traditional MOLP to the interval objective function case [e.g. (Bitran 1980; Ida 1999, 2000a, b, 2005; Inuiguchi and Sakawa 1996; Steuer 1981; Wang and Wang 2001a, b; Oliveira et al. 2014)]. Bitran suggested two types of efficient solutions for the interval MOLP: a "necessarily efficient" if it is efficient for all objective function coefficient vectors within their admissible range of variation (see the vertex with a bold circle obtained with the gradient cones of the two objective functions illustrated in Fig. 1); a "possibly efficient" if it is efficient for at least one of the given objective function coefficient vectors within their admissible range of variation (Bitran 1980). When compared to the "possibly efficient"

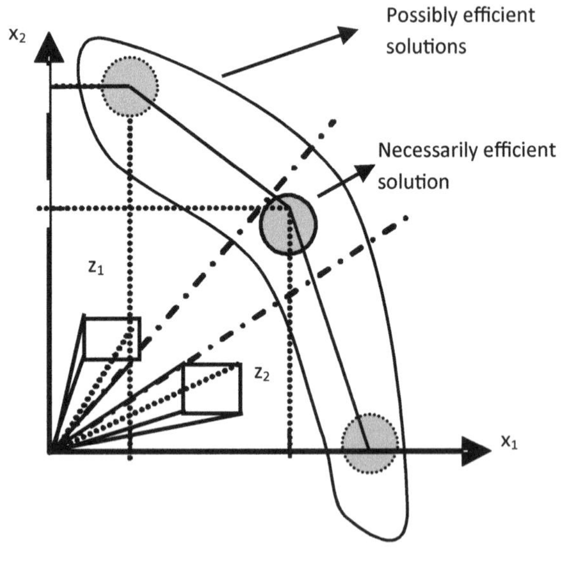

Fig. 1 Necessarily and possibly efficient solutions (Oliveira and Antunes 2007)

solutions, the "necessarily efficient" solutions are the most robust (Ida 1999). Although this type of approach allows enumerating all possibly efficient solutions and/or all necessarily efficient solutions, the computational burden involved can be significant. Another issue is that with a large set of solutions, in many cases with just slight differences among the objective function values, the decision problem becomes even more complex (Antunes and Clímaco 2000).

Interactive approaches have also been considered to obtain solutions to the MOLP models with interval coefficients in the whole model. Urli and Nadeau have proposed an interactive algorithm that does not allow the DM to take into account the worst case and the best case "scenarios" (Urli and Nadeau 1992). In the algorithm proposed by Oliveira and Antunes the procedures involved provide a global view of the solutions in the best and worst case coefficients scenario and allow performing the search of new solutions according with the achievement rates of the objective functions, both regarding the upper and lower bounds (Oliveira and Antunes 2009). The main aim is to identify the solutions associated with the interval objective function values which are closer to their corresponding interval ideal solutions. With this approach it is also possible to find solutions with non-dominance relations regarding the achievement rates of the upper and lower bounds of the objective functions, considering interval coefficients in the whole model.

2.3 MOITP Formulation

The conventional MOTP problem can be formulated, without loss of generality, as:

$$
\begin{aligned}
&\text{Minimize } Z^k = \sum_{i=1}^{m}\sum_{j=1}^{n} c_{ij}^k x_{ij}, \quad \text{where } k = 1, \ldots, p. \\
&\text{subject to} \\
&\sum_{j=1}^{n} x_{ij} = a_i, \quad i = 1, \ldots, m. \\
&\sum_{i=1}^{m} x_{ij} = b_j, \quad j = 1, \ldots, n \\
&x_{ij} \geq 0, \quad i = 1, \ldots, m, \quad j = 1, \ldots, n. \\
&\text{with } \sum_{i=1}^{m} a_i = \sum_{j=1}^{n} b_j,
\end{aligned}
\tag{8}
$$

where x_{ij} is the decision variable which refers to product quantity that has to be transported from supply point i to demand point j; c_{ij}^k, k = 1, …, p, denotes the unit transportation cost from ith supply point to jth demand; a_i, i = 1, …, m, represents the ith supply quantity; b_j, j = 1, …, m, represents the jth demand quantity.

Three major cases can occur in MOITP problems (Das et al. 1999):

1. The coefficients of the objective functions are given in the form of interval values, whereas source and destination parameters are deterministic.
2. The source and destination parameters are in the form of interval values but the objective functions' coefficients are deterministic.
3. All the coefficients and parameters are in the form of interval values.

If all the coefficients and parameters are provided in the interval form, the MOITP can be formulated, without loss of generality, as the problem of minimizing p interval-valued objective functions, with interval source and interval destination parameters, as follows (Das et al. 1999; Sengupta and Pal 2009):

$$\text{Minimize } Z^k = \sum_{i=1}^{m}\sum_{j=1}^{n}\left[c^k_{Lij}, c^k_{Uij}\right]x_{ij}, \quad \text{where } k = 1,\ldots,p.$$

subject to

$$\begin{aligned} &\sum_{j=1}^{n} x_{ij} = [a_{Li}, a_{Ui}], \quad i = 1,\ldots,m.\\ &\sum_{i=1}^{m} x_{ij} = [b_{Lj}, b_{Uj}], \quad j = 1,\ldots,n.\\ &x_{ij} \geq 0, \quad i = 1, \ldots, m,\ j = 1, \ldots, n.\\ &\text{with } \sum_{i=1}^{m} a_{Li} = \sum_{j=1}^{n} b_{Lj} \text{ and } \sum_{i=1}^{m} a_{Ui} = \sum_{j=1}^{n} b_{Uj}\\ &\text{or with } \sum_{i=1}^{m}\frac{1}{2}(a_{Li} + a_{Ui}) = \sum_{j=1}^{n}\frac{1}{2}(b_{Lj} + b_{Uj}), \end{aligned} \tag{9}$$

where x_{ij} is the decision variable which refers to product quantity that has to be transported from supply point i to demand point j; $\left[c^k_{Lij}, c^k_{Uij}\right]$, k = 1, …, p, denotes the unit transportation cost from ith supply point to jth demand comprised between c^k_{Lij} and c^k_{Uij}; $[a_{Li}, a_{Ui}]$, i = 1, …, m, represents the ith supply quantity within a_{Li} and a_{Ui}; $[b_{Lj}, b_{Uj}]$, j = 1, …, m, represents the jth demand quantity located between b_{Lj} and b_{Uj}.

3 Solution Methods for MOITP

Abd El-Wahed and Lee classify the solution approaches for MOTP into four categories: interactive approaches [see e.g. (Climaco et al. 1993; Ringuest and Rinks 1987)], non-interactive approaches [see e.g. (Aneja and Nair 1979)], goal programming (GP) approaches [see e.g. (Hemaida and Kwak 1994)] and fuzzy programming approaches [see e.g. (Abd El-Wahed and Lee 2006; Abd El-Wahed 2001; Li and Lai 2000)]. The interactive approaches may present several limitations: the convergence to an efficient solution depends on the DM's reasoning and consistency; unless the interactive method becomes more flexible to facilitate the enumeration and evaluation of the set of efficient solutions in large scale problems, the procedure bears a high computational burden. However, with this sort of approaches the search direction of efficient solutions is controlled by the DM allowing him/her to reach an efficient solution consistent with his/her preferences. The non-interactive approaches depend on the generation set of efficient solutions, and the choice of the preferred compromise solution out of this set is then required from the DM. Therefore, the solution search process may require a large computational burden and the DM may be facing additional difficulties in assessing the tradeoffs between the different solutions. The GP approach allows obtaining a compromise solution according to the established goal levels (aspiration levels). Nevertheless, it is often difficult for the DM to decide the desired aspiration levels for the goals, eventually leading to non-efficient solutions; the choice of the weights in the formulation of the GP problem may also lead to non-satisfying results; erroneous conclusions can be obtained if the achievement function is not correctly formulated; the GP formulation changes the traditional mathematical form of the MOTP problem. Finally, the fuzzy programming approaches are based on the use of fuzzy set theory for solving the MOTP by means of an interactive procedure. The use of fuzzy set theory in solving such MOTP changes the standard form of the TP. Moreover, the computation of an efficient solution is not guaranteed in certain conditions [see e.g. (Li and Lai 2000)]. Although this section is devoted to the MOITP, our analysis will also encompass methods used in the framework of interval TP (ITP). Therefore, by extending the previous classification of solution approaches to MOITP and ITP, it might be concluded that the approaches mainly used are broadly classified into three categories: the composite approach (i.e., they are the result of the combination of two or more approaches), the fuzzy approach (usually used for obtaining a final solution) and non-fuzzy approach. In the framework of the composite approach, Abd El-Wahed and Lee (2006) presented an interactive, fuzzy and goal programming approach to determine the preferred compromise solution for the MOTP. The suggested approach considers the imprecise nature of the input data by implementing the minimum operator, also assuming that each objective function has a fuzzy goal. The approach focuses on minimizing the worst upper bound to obtain an efficient solution which is close to the best lower bound of each objective function. The solution procedure controls the search direction by updating both the membership values and the aspiration levels. An important characteristic of the approach is that the decision-maker's (DM's) role is concentrated only in evaluating the efficient solution to limit the influences of his/her incomplete knowledge about the problem domain. The approach controls the search direction by updating both upper bounds and aspiration level of each objective function. In the context of the fuzzy approach, Das et al. suggested a procedure that starts by transforming the MOITP (see problem (8)) into a classical MOTP, where the objectives (the upper bound and midpoint of the interval objective functions) are minimized and the constraints with interval source and destination parameters have been converted into deterministic ones (Das et al. 1999). Finally, the equivalent transformed problem has been solved by the fuzzy programming technique. Sengupta and Pal proposed a methodology for solving an ITP where multiple penalty factors are involved, reflecting the DM's pessimist or optimistic bias in achieving the compromise solution by means of a fuzzy oriented method (Sengupta and Pal 2009). Panda and Das presented a two-vehicle cost varying interval transportation model (TVCVITM) where the cost varies due to the capacity of the vehicles and to the quantity transported (Panda and Das 2013). The source and destination parameters are considered as intervals. Initially, depending on the cost of the vehicles the interval coefficients of the objective function are specified. Then, the problem is converted into a classical ITP. Finally, this model is converted into a bi-objective TP, where the upper bound and the midpoint of the objective function are minimized. The solution to this bi-objective model is obtained with a fuzzy programming technique. Nagarajan et al. suggested a solution procedure for the MOITP problem under stochastic environment using fuzzy programming approach. All source availability, destination demand and conveyance capacities have been taken as stochastic intervals for each criterion. Expectation of a random variable has been used to transform the problem into a classical MOTP where the objectives which are the upper bound and midpoint of the interval objective functions are minimized (Nagarajan et al. 2014). In the context of non-fuzzy approach, Pandian and Natarajan (2010) suggested a new method (the separation method based on the zero point method) for finding an optimal solution for the integer TP where transportation cost, supply and demand are given as interval values. The proposed method has been developed without using the midpoint and width of the interval in the objective function. Joshi and Gupta investigated the transportation problem with fractional objective function when the demand and supply quantities are varying. The method computes the lower and upper bounds of the total fractional transportation cost when the supply and demand quantities are varying. A set of two-level transportation problems is transformed into one-level mathematical programs to obtain the objective value (Joshi and Gupta 2011). Pandian and Anuradha (2011) applied a split and bound method for finding an optimal solution to a fully integer interval TP with additional impurity constraints. This method has been developed without considering the midpoint and width of the intervals and is based on the floating point method [see Pandian and Anuradha (2011b)]. Rakocevic and Dragasevic (2011) presented a parametric TP, in which the coefficients of the objective function depend on a parameter. The main aim of this proposal is to assess the impact on the optimal solution of changes in the coefficients of the objective function, i.e. to find the range of variation of the coefficients of the objective function without affecting the optimal solution obtained. Roy and Mahapatra (2011) dealt with the interval coefficients to the multiobjective stochastic TP, involving an inequality type of constraints in which all parameters (supply and demand) are lognormal random variables and the coefficients of the objective functions are interval numbers. The minimization MOITP is also converted into a bi-objective problem using the order relations which represent the DM's preferences between the interval costs. These interval costs have been defined by the upper and lower bounds and corresponding midpoint and width. The proposed probabilistic constraints are firstly converted into an equivalent deterministic constraint by means of the chance constrained programming technique. Then, a surrogate problem has been solved by the weighted sum method. Guzel et al. (2012) transformed a fractional TP with interval coefficients into a classical TP. Two solution procedures are then proposed for the interval fractional TP. One of them is based on a Taylor series approximation and the other one is based on interval arithmetic. Kavitha and Pandian (2012) presented the sensitivity analysis of supply and demand parameters of an ITP. The method is aimed at determining the ranges of supply and demand parameters in an interval transportation problem such that its optimal basis is invariant. The upper-lower method is used for finding a critical region of the supply and demand parameters at which any change inside the ranges of the region does not affect the optimal basis, while any change outside their ranges will affect the optimal basis. Dalman et al. (2013) considered the Indefinite Quadratic Interval TP (IQITP) in which all the parameters i.e. cost and risk coefficients of the objective function, supply and demand quantities are expressed as intervals. Firstly, a feasible initial point is determined within the Northwest Corner method by means of expressing all the interval parameters as left and right limits. Then the objective function is linearized by using first order Taylor series expansion about the feasible initial point. Thus IQITP is transformed into a traditional LP problem. Then an iterative procedure is presented in such a way that the optimal solution of the last LP problem is selected as the point from which the objective will be expanded into its first order Taylor series in the next iteration step. The stopping criterion of the proposed procedure is obtaining the same point for the last two iteration steps. Fegade et al. (2013) considered TP with and without budgetary constraints, where demand and budget are imprecise. An interval-point method for finding an optimal solution for transportation problems is proposed and compared with zero suffix method. Panda and Das (2014) also presented the two-vehicle cost varying transportation problem as a bi-level mathematical programming model. The Northwest Corner rule is used for determining the initial basic feasible solution and then the unit transportation cost (which varies in each iteration) is established according to the choice of vehicles. These authors also concluded that the two-vehicle cost varying transportation model provides more efficient results than the single objective cost varying transportation problem.

3.1 Advantages and Disadvantages of Solution Methods

From the different methodologies briefly presented it might be concluded that the combination of two or more approaches may reduce some or all of the shortcomings of each individual approach. Another issue refers to the fact that the majority of the approaches herein reviewed transforms the original MOITP/TP into a surrogate crisp problem and then apply some algorithm to solve it (see Table 1). As a result, the uncertainty of the intervals may get lost to a certain extent. In fact, in view of interval order relations, the satisfactory solutions attained sometimes allow obtaining contradictory results. The example suggested by Das et al. and used by Sengupta and Pal helps illustrating this issue (Oliveira and Antunes 2009; Das et al. 1999).

Consider the following example:

$$\text{Minimize } Z^k = \sum_{i=1}^{3}\sum_{j=1}^{4}\left[c^k_{Lij}, c^k_{Uij}\right] x_{ij}, \quad \text{where } k = 1, 2.$$

subject to

$$\sum_{j=1}^{4} x_{1j} = [7, 9], \quad \sum_{j=1}^{4} x_{2j} = [17, 21], \sum_{j=1}^{4} x_{3j} = [16, 18],$$

$$\sum_{i=1}^{3} x_{i1} = [10, 12], \sum_{i=1}^{3} x_{i2} = [2, 4], \sum_{i=1}^{3} x_{i3} = [13, 15], \sum_{i=1}^{3} x_{i4} = [15, 17],$$

$$x_{ij} \geq 0, \ i = 1, 2, 3, \ j = 1, 2, 3, 4.$$

$$\text{where } c^k_{ij} = \left[c^k_{Lij}, c^k_{Uij}\right], \ k = 1, 2,$$

$$c^1_{ij} = \begin{bmatrix} [1, & 2] & [1, & 3] & [5, & 9] & [4, & 8] \\ [1, & 2] & [7, & 10] & [2, & 6] & [3, & 5] \\ [7, & 9] & [7, & 11] & [3, & 5] & [5, & 7] \end{bmatrix} \text{ and}$$

$$c^2_{ij} = \begin{bmatrix} [3, & 5] & [2, & 6] & [2, & 4] & [1, & 5] \\ [4, & 6] & [7, & 9] & [7, & 10] & [9, & 11] \\ [4, & 8] & [1, & 3] & [3, & 6] & [1, & 2] \end{bmatrix}.$$

With the fuzzy programming technique suggested by Das et al. (1999) the following Pareto optimal solution is attained: $x_{12} = 2$, $x_{14} = 5$, $x_{21} = 10$, $x_{23} = 6.01$, $x_{24} = 0.99$, $x_{33} = 6.97$, $x_{34} = 9.01$, with $Z^1 = [113, 204.9]$, $w[Z^1] = 91.9$, $m[Z^1] = 158.95$ and $Z^2 = [129.89, 227.86]$, $w[Z^2] = 97.97$, $m[Z^2] = 178.85$. By means of the fuzzy oriented method proposed by Sengupta and Pal (2009) the following solution is obtained, considering a higher importance of the first objective function and with a more certain result—$x_{12} = 2$, $x_{13} = 5$, $x_{21} = 10$, $x_{24} = 7$, $x_{33} = 8$, $x_{34} = 8$, with $Z^1 = [122, 202]$, $w[Z^1] = 80$, $m[Z^1] = 162$ and $Z^2 = [149, 233]$, $w[Z^2] = 84$, $m[Z^2] = 191$.

Let $Z^{1D} = [113, 204.9]$ and Let $Z^{2D} = [129.89, 227.86]$ be the interval solutions obtained by Das et al. (1999) and $Z^{1S} = [122, 202]$ and $Z^{2S} = [149, 233]$ be the interval solutions reached by Sengupta and Pal (2009). If we compare these interval objective function values using the interval order relations discussed in Sect. 2.1, it can be concluded that:

Table 1 Approaches for obtaining solutions for the MOITP/ITP

Approach	Advantages	Disadvantages
Composite (interactive, fuzzy and GP) (Abd El-Wahed and Lee 2006)	Solves other MOLP problems. Combines fuzzy, GP and interactive approaches, overcoming the conflict among the objectives and the problems of determining the aspiration levels via GP. Provides a preferred compromise solution from the DM's perspective. Determines appropriate aspiration levels of the objective functions	May lead to computation related problems
Fuzzy (Das et al. 1999)	Leads to k non-dominated solutions and one optimal compromise solution with k objective functions	Transforms the MOITP problem into a classical MOTP
Fuzzy (Sengupta and Pal 2009)	Reflects the DM's preferences changing from pessimism to optimism. Requires few steps to yield the compromise solution. Allows flexibility regarding the way the DM reaches his/her desired satisfactory solution	Transforms the problem into a crisp surrogate
Fuzzy (Panda and Das 2013)	Uses two-vehicle cost varying TP	Considers a single objective function
Stochastic/fuzzy (Nagarajan et al. 2014)	Leads to k non-dominated solutions and one optimal compromise solution with k objective functions	Transforms the MOITP problem into a classical MOTP
Non-fuzzy (separation method) (Pandian and Natarajan 2010)	Considers the decision variables are also given as intervals. Uses the separation method without using the midpoint and width of the interval in the objective function of the fully interval TP which is a non-fuzzy method. Extends the approach to fuzzy TP	Considers a single objective function. Transforms the problem into a crisp surrogate
Non-fuzzy (non-interactive) (Joshi and Gupta 2011)	Calculates the total cost bounds of the transportation problem, where at least either the supply or demand is varying	Considers a single objective function. Requires fine-tuning the production strategy or warehouse stocking level for reducing the total transportation cost. The largest total transportation cost may not occur at the highest total quantity shipped

(continued)

Table 1 (continued)

Approach	Advantages	Disadvantages
Non-fuzzy (floating point method) (Pandian and Anuradha 2011a, b)	Considers the decision variables are also given as intervals. Uses the floating point method without using the midpoint and width of the interval in the objective function of the fully interval TP which is a non-fuzzy method. Extends the approach to fuzzy TP	Considers a single objective function. Transforms the problem into a crisp surrogate
Non-fuzzy (parametric) (Rakocevic and Dragasevic 2011)	Allows determining the behavior of the system if the parameters change within the given limits, as well as to set parameters so that, with the optimal solution determined, the greatest possible success is achieved, without significant distortion of the structure of the optimal solution	Considers a single objective function
Non-fuzzy (stochastic) (Roy and Mahapatra 2011)	Reflects the DM's pessimistic or optimistic bias in achieving a non-inferior solution. Facilitates the DM's choice of the desired satisfactory solution by the suitable arrangement of the objective function weights. Only a few steps are required to obtain the non-inferior solution	Transforms the MOITP problem into a classical MOTP
Non-fuzzy (Taylor series approximation and Interval arithmetic) (Guzel et al. 2012)	Introduces two solution procedures for the fractional ITP	Transforms a fractional TP with interval coefficients into a classical TP. Considers a single objective function
Non-fuzzy (sensitivity analysis) (Kavitha and Pandian 2012)	Extends the approach to fuzzy TP. Serves as an important tool for helping DM taking appropriate decisions when handling various types of logistic problems having imprecise parameters	Considers a single objective function.
Non-fuzzy (iterative) (Dalman et al. 2013)	Incorporates the total cost of damaged goods into the problem. Enables the DM to consider tolerances for the model parameters in a more natural and direct way. Uses quadratic functions	Considers a single objective. Transforms the IQITP into a traditional LP problem
Non-fuzzy (Interval-point method) (Fegade et al. 2013)	Enables the DM to evaluate TP with budgetary constraints	Considers a single objective function. Transforms the problem into a crisp surrogate
Non-fuzzy (Non-interactive) (Panda and Das 2014)	Uses two-vehicle cost varying TP	Considers a single objective function

1. According to Ishibuchi and Tanaka (1990) $Z^{1D} \leq_{MW} Z^{1S}$, since $m[Z^{1D}] \leq m[Z^{1S}]$ and $w[Z^{1D}] \geq w[Z^{1S}]$; $Z^{2D} \leq_{MW} Z^{2S}$, since $m[Z^{2D}] \leq m[Z^{2S}]$ and $w[Z^{2D}] \geq w[Z^{2S}]$;
2. According to Ishibuchi and Tanaka (1990) $Z^{1D} \leq_{LM} Z^{1S}$, since the "lower bound of Z^{1D}" ≤ "lower bound of Z^{1S}" and $m[Z^{1D}] \leq m[Z^{1S}]$; $Z^{2D} \leq_{LM} Z^{2S}$, since the "lower bound of Z^{2D}" ≤ "lower bound of Z^{2S}" and $m[Z^{2D}] \leq m[Z^{2S}]$;
3. Using the acceptability index developed by Sengupta and Pal (2000) $Z^{ID} \prec Z^{IS}$, since

$$Z^{1D} \prec = \frac{\left(m\left[Z^{1S}\right] - m\left[Z^{1D}\right]\right)}{\left(\frac{w\left[Z^{1S}\right]}{2} + \frac{w\left[Z^{1D}\right]}{2}\right)} = \frac{(162 - 158.95)}{(40 + 45.95)} = 0.035485748 > 0 \quad \text{and}$$

$$Z^{2D} \prec Z^{2S}, \text{since}\, Z^{2D} \prec = \frac{\left(m\left[Z^{2S}\right] - m\left[Z^{2D}\right]\right)}{\left(\frac{w\left[Z^{2S}\right]}{2} + \frac{w\left[Z^{2D}\right]}{2}\right)} = \frac{(191 - 178.85)}{(42 + 48.985)} = 0.133263725 > 0.$$

In a minimization problem, if both $\mathcal{A}\left(Z^{1D} \prec Z^{1S}\right) > 0$ and $\mathcal{A}\left(Z^{2D} \prec Z^{2S}\right) > 0$ it should be concluded that Z^{1D} is better than Z^{1S} and Z^{2D} is better than Z^{2S}, which allows concluding that from this point of view the solution proposed by Sengupta and Pal (2009) is dominated by the solution obtained by Das et al. (1999). In fact, all the interval order relations indicate the same results.

Table 1 provides a brief overview of the main advantages and drawbacks found in the approaches previously discussed.

4 Conclusions

The mathematical formulation of the traditional MOTP assumes that the coefficients of the objective functions and the supply and demand values are considered as crisp values. However, the unit transportation costs may vary and the supply and demand may also change. In this context, interval programming is one of the approaches used to handle uncertainty in mathematical programming models, which entails some interesting characteristics. In contrast to stochastic programming or to fuzzy programming which start with the specification or the assumption of probabilistic distributions and possibilistic distributions, respectively, and to robustness optimization techniques which inherently consider a max-min formulation (i.e., worst-case), interval programming only requires information about the range of variation of some (or all) of the parameters. In the framework of the critical assessment of the interval programming techniques for solving MOITP revisited in this study, it might be concluded that the combination of two or more approaches in the solution methods applied to MOITP/ITP may reduce the limitations of each individual approach. On the other hand, in general, the approaches herein reviewed transform the original MOITP/TP into a crisp model and then use an algorithm to solve it. Consequently, the uncertainty of the intervals becomes lost to a certain extent. Finally, as it was shown, the satisfactory solutions obtained may lead to contradictory results from the point of view of interval order relations. Therefore, new ways to model the MOITP/TP should be proposed in order to overcome this expected drawback.

Acknowledgments This work has been supported by the Fundação para a Ciência e a Tecnologia (FCT) under project grant UID/MULTI/00308/2013.

References

Aneja, Y. P., & Nair, K. P. K. (1979). Bicriteria transportation problems. *Management Science, 25* (1), 73–78.

Antunes, C. H., Clímaco, J. (2000). Decision aid in the optimization of the interval objective function. In S. H. Zanakis, G. Doukidis & C. Zopounidis (Eds.), *Decision making: Recent developments and worldwide applications* (Vol. 45, pp. 251–261). Kluwer Academic Publishers, Applied Optimization.

Bitran, G. R. (1980). Linear multiobjective problems with interval coefficients. *Management Science, 26*(7), 694–706.

Chanas, S., & Kuchta, D. (1996). Multiobjective programming in optimization of interval objective functions—a generalized approach. *European Journal of Operational Research, 94* (3), 594–598.

Climaco, J. N., Antunes, C. H., & Alves, M. J. (1993). Interactive decision support for multiobjective transportation problems. *European Journal of Operational Research, 65*(1), 58–67.

Dalman, H., Koçken, H., & Sivri, M. (2013). A solution proposal to indefinite quadratic interval transportation problem. *New Trends in Mathematical Sciences, 1*(2), 7–12.

Das, S. K., Goswami, A., & Alam, S. S. (1999). Multiobjective transportation problem with interval cost, source and destination parameters. *European Journal of Operational Research, 117*, 100–112.

El-Wahed, W. F. (2001). A multi-objective transportation problem under fuzziness. *Fuzzy Sets and Systems, 117*(1), 27–33.

El-Wahed, W. F., & Lee, S. M. (2006). Interactive fuzzy goal programming for multi-objective transportation problems. *Omega, 34*, 158–166.

Fegade, M. R., Jadhav, V. A., Doke, D. M., & Irpenwar, B. M. (2013). Comparative study of multi-objective transportation problems with and without budgetary constraints. *International Journal of Statistika and Mathematika., 6*(1), 18–21.

Guzel, N., Emiroglu, Y., Tapci, F., Guler, C., & Syvry, M. (2012). A solution proposal to the interval fractional transportation problem. *Applied Mathematics & Information Sciences*, *6*(3), 567–571.

Hemaida, R. S., & Kwak, N. K. (1994). A linear goal programming model for trans-shipment problems with flexible supply and demand constraints. *Journal of Operational Research Society, 45*(2), 215–224.

Ida, M. (1999). Necessary efficient test in interval multiobjective linear programming. In *Proceedings of the Eighth International Fuzzy Systems Association World Congress* (pp. 500–504).

Ida, M. (2000a). Efficient solution generation for multiple objective linear programming and uncertain coefficients. In *Proceedings of the 8th Bellman Continuum* (pp. 132–136).

Ida, M. (2000b) Interval multiobjective programming and mobile robot path planning. In M. Mohammadian (Eds.), *New frontiers in computational intelligence and its applications* (pp. 313–322). IOS Press.
Ida, M. (2005). Efficient solution generation for multiple objective linear programming based on extreme ray generation method. *European Journal of Operational Research, 160*(1), 242–251.
Inuiguchi, M., & Kume, Y. (1994). Minimax regret in linear programming problems with an interval objective function. In G. H. Tzeng, H. F. Wang, U. P. Wen, & P. L. Yu (Eds.), *Multicriteria decision making* (pp. 65–74). New York: Springer.
Inuiguchi, M., & Sakawa, M. (1995). Minimax regret solution to linear programming problems with an interval objective function. *European Journal of Operational Research, 86*(3), 526–536.
Inuiguchi, M., & Sakawa, M. (1996). Possible and necessary efficiency in possibilistic multiobjective linear programming problems and possible efficiency test. *Fuzzy Sets and Systems, 78*(2), 241–321.
Ishibuchi, H., & Tanaka, H. (1990). Multiobjective programming in optimization of the interval objective function. *European Journal of Operational Research, 48*(2), 219–225.
Joshi, V. D., & Gupta, N. (2011). Linear fractional transportation problem with varying demand and supply. *Le matematiche, 66*(2), 3–12.
Kavitha, K., & Pandian, P. (2012). Sensitivity analysis of supply and demand in a fully interval transportation problem. *International Journal of Engineering Research and Applications, 2*(3), 1900–1910.
Li, L., & Lai, K. K. (2000). A fuzzy approach to the multiobjective transportation problem. *Computers & Operational Research, 27*(1), 43–57.
Moore, R. E. (1966). *Interval analysis*. New Jersey: Prentice-Hall.
Nagarajan, A., Jeyaraman, K., & Prabha, Sk. (2014). Multi_objective solid transportation problem with interval cost in source and demand parameters. *International Journal of Computer & Organization Trends, 8*(1), 33–41.
Oliveira, C., & Antunes, C. H. (2007). Multiple objective linear programming models with interval coefficients–an illustrated overview. *European Journal of Operational Research, 181*(3), 1434–1463.
Oliveira, C., & Antunes, C. H. (2009). An interactive method of tackling uncertainty in interval multiple objective linear programming. *Journal of Mathematical Sciences, 161*(6), 854–866.
Oliveira, C., Antunes, C. H., & Barrico, C. (2014). An enumerative algorithm for computing all possibly optimal solutions to an interval LP. *TOP—Journal of the Spanish Statistical and Operations Research Society, 22*, 530–542.
Panda, A., & Das, C. B. (2013). Cost varying interval transportation problem under two vehicle. *Journal of New Results in Science, 3*, 19–37.
Panda, A., & Das, C. B. (2014). 2-vehicle cost varying transportation problem. *Journal of Uncertain Systems, 8*(1), 44–57.
Pandian, P., & Anuradha, D. (2011a). Solving interval transportation problems with additional impurity constraints. *Journal of Physical Sciences, 15*, 103–112.
Pandian, P., & Anuradha, D. (2011b). Floating point method for solving transportation problems with additional constraints. *International Mathematical Forum, 6*(40), 1983–1992.
Pandian, P., & Natarajan, G. (2010). A new method for finding an optimal solution of fully interval integer transportation problems. *Applied Mathematical Sciences, 4*(37), 1819–1830.
Rakocevic, S., & Dragasevic, Z. (2011). Parametric programming for the transportation problem. In *The 15th International Research/Expert Conference "Trends in the Development of Machinery and Associated Technology" TMT 2011* (pp. 925–928). Prague, Czech Republic, 12–18 Sept 2011.
Ringuest, J. L., & Rinks, D. B. (1987). Interactive solutions for the linear multiobjective transportation problems. *European Journal of Operational Research, 32*(1), 96–106.
Roy, S. K., & Mahapatra, D. R. (2011). Multi-objective interval-valued transportation probabilistic problem involving log-normal. *International Journal of Mathematics And Scientific Computing, 1*(2), 14–21.

Sengupta, A., & Pal, T. K. (2000). On comparing interval numbers. *European Journal of Operational Research, 127*(1), 28–43.
Sengupta, A., & Pal, T. K. (2009). Transportation problem with multiple penalty factors. In A. Sengupta & T. K. Pal (Eds.), *Fuzzy preference ordering of interval numbers in decision problems* (pp. 121–137). Berlin Heidelberg: Springer-Verlag.
Steuer, R. E. (1981). Algorithms for linear programming problems with interval objective function coefficients. *Mathematics of Operations Research, 6*(3), 333–348.
Urli, B., & Nadeau, R. (1992). An interactive method to multiobjective linear programming problems with interval coefficients. *INFOR, 30*(2), 127–137.
Wang, M. L., & Wang, H. F. (2001a). Decision analysis of the interval–valued multiobjective linear programming problems. In M. Köksalan & S. Zionts (Eds.), *Lecture notes in economics and mathematical systems, multiple criteria decision making in the new millennium* (Vol. 507, pp. 210–218). Springer.
Wang, M. L., & Wang, H. F. (2001b). Interval analysis of a fuzzy multiobjective linear programming. *International Journal of Fuzzy Systems, 34*, 558–568.

Lean Management and Supply Chain Management: Common Practices

Jordi Olivella Nadal

Abstract In Operations Management, different approaches to make a company successful are applied. Which practices are part of which approaches is not always clear. This work is devoted to a particular comparison between two specific approaches, Lean and Supply Chain Management (SCM). The comparison is presented as an example of the relation between two sets of practices and is also interesting in itself. Starting from the concepts and practices included in Lean and SCM according to the literature, the common practices are obtained. Lean Supply Chain is also taken into account. A critical point in common between Lean and SCM, Value Stream Management, is also highlighted.

1 Introduction

Students and researchers of Operations Management have to confront a range of approaches supposedly able to make a company successful. The list is long: Total Quality Management (TQM), Lean Management (hereafter called Lean), World Class Manufacturing, Six Sigma, Supply Chain (SC), Supply Chain Management (SCM), and so on. The immediate question is what the best option is.

As no approach is clearly predominant, we can think that it depends on the characteristics of each case. But a closer analysis shows that we cannot give a conclusive answer because there is not a single Total Quality or Lean set of practices, for example. Very different practices are developed by organizations that claim to follow these approaches. In fact, most of the tools and practices are included in several or all the approaches.

When someone states that a Lean program is being implemented, we could think that they are following a program similar to the one the best known lean companies

J. Olivella Nadal (✉)
Department of Management, Institute of Industrial and Control Engineering, Universitat Politècnica de Catalunya, Av. Diagonal, 647, 08028 Barcelona, Spain
e-mail: jorge.olivella@upc.edu

A.P. Barbosa-Póvoa et al. (eds.), *Optimization and Decision Support Systems for Supply Chains*, Lecture Notes in Logistics,
DOI 10.1007/978-3-319-42421-7_8

apply and some books describe. However, the Lean program that they are implementing can be more similar to some TQM or Six Sigma implementations than to other Lean implementations.

The uncertainty about what the content of a Lean or a TQM implementation really is, probably less damaging in practice that it could seem. A real implementation is an adaptation of the same principles and tools to the context of the company involved. Companies are not worried about whether they are following or not a "pure" version of some theory. Lean or TQM, for example, can be useful for them as an inspiration and also as a communication tool, even if they do not imply a specific program. In any case, confusion regarding these concepts can raise doubts.

This work is devoted to a particular comparison between two approaches, Lean and SCM. It is presented as an example of the relation between two well-known sets of practices and is also interesting in itself. To make the comparison, two lists of principles and concepts of Lean and SCM are considered. The common practices are presented and a special emphasis is given to a mixed approach, Lean Supply Chain (LSC), and to a critical common practice of Lean and SCM, Value Stream Management (VSM).

In the academic field, the different combinations of practices that are included in Lean implementations have been called "lean bundles". Naturally, this expression is not used by managers or consultants promoting Lean, as asking a company to implement something called "bundle" would not be commercial. But it is a useful concept for observers. In a set of analysed examples of Lean implementations, only three practices were found in all of them (Shah 2003): Just in Time (JIT), pull and quick changeover. Although they are of capital importance, the number of coincidences between all "lean bundles" is small. The topic of what the most important elements of Lean are is addressed in Sect. 2.

Many doubts arise when dealing with SC or SCM. The concept of Supply Chain comprises the different elements of the network, from raw material to the final customer. Beyond this general coincidence, many different definitions have been suggested. Some include objectives of SC and SCM as "customer satisfaction" or "maximizing profitability" (Corominas 2013). Common points between SCM and Lean can be potentially numerous when the objectives of SCM are considered. Coincidences between Lean and SCM are discussed in Sect. 3.

LSC is the application of Lean principles to the whole supply chain, that is to say, the application of Lean approach to a wider approach. LSC is an extension of Lean and will not necessarily consider all the elements of the SCM. Section 4 deals with LSC.

A central concept both in Lean and SCM is VSM. Regarding Lean, the concepts of Value Stream and Flow Management define the lean enterprise (Womack and Jones 1994). The identification of wastes and the focus on customer value cannot be obtained with a functional or departmental vision. Similarly, SC integration, that is to say, integrated management of the flow through improved communication, partnerships, alliances, and cooperation, is one of the central topics in Supply Chain literature (Power 2005). For this reason, Sect. 5 is devoted to VSM. Finally, Sect. 6 is devoted to the conclusions.

2 What is Lean Really?

We could define Lean as the set of principles, practices, and tools that the companies that claim to be Lean usually apply, inspired by the practices of Toyota. Such a definition establishes the scope but not the content of the elements included. The complete list of practices and tools used to support them is too long to consider it as an effective reference. The challenge is to obtain a limited number of ideas that give sense to the whole approach.

The literature is rich in analysing what the essence of Lean is. Often the objectives of "removing waste", that is to say, eliminating dispensable activities, and "focusing on customer value" have been considered to be central in Lean approach. However, to define objectives and to reach them is not the same. It has been reported that the success of a Lean implementation depends on the systematic application of the scientific approach in the day to day activities (Spear and Bowen 1999). A different vision of the Lean system emphasizes its Operations Management aspects. The diverse aspects are reflected in the 14 principles of Toyota (Liker 2003).

To compare Lean with other approaches such as SCM, it is necessary to focus on the practices that are commonly related to Lean approach. Lean implementation involves different aspects of a company activity. We can distinguish between production methods, work practices, and cultural aspects. The following production methods and tools have been considered to be central to Lean (Belekoukias et al. 2014):

- Just in Time. A manufacturing philosophy focusing on the elimination of waste (non-value added activities) in the manufacturing process by the most timely sequencing of operations (Gass and Fu 2013).
- Total Productive Maintenance (TPM). TPM is commonly associated with autonomous and planned maintenance activities and includes other activities that help to improve equipment effectiveness over the entire life of the equipment (Swamidass 2000).
- Autonomation. Autonomation means the autonomous control of quality and quantity. The initial idea was that every worker is personally responsible for the quality of the part or product that he/she produced. Quite often the inspection is performed automatically (Swamidass 2000).
- Value stream focus. Value stream includes the complete value-adding process, from the conception of requirement back through to raw material source and back again to the consumer's receipt of the product. Value stream focus is critical to remove wastes. Value Stream Mapping (VSM) is a tool that provides visibility along the value stream (Hines and Rich 1997).
- Continuous improvement. Continuous improvement or Kaizen is a constant endeavour to expose and eliminate problems at the root level (Swamidass 2000).

The implementation of Lean and, in particular, the production approaches mentioned above requires that day to day work practices are appropriately adapted. A description of Lean that would not include these aspects would give a false image of what Lean is. These practices can be summarized in the following points (Olivella et al. 2008):

- Standardization, Discipline, and Control. Standardization is an essential principle of Lean, encompassing both the sequence of tasks to be done by each worker and how those tasks are done. Discipline and close control are also indispensable for Lean.
- Continuing Training and Learning. Training and learning are critical to the implementation of Lean. Workers obtain knowledge from previous, initial and continuous training and, most importantly, from experience.
- Team-based Organization. Teamwork refers to joint, shared work. In organizations based on work teams, responsibilities—particularly, workloads—are assigned to the teams.
- Participation and Empowerment. Lean requires a leadership style in which hierarchical superiority is deemphasised, and which includes a system of suggestions and planned discussion.
- Common Values. Lean involves active engagement of all the staff that cannot be obtained through disciplinary procedures but from worker commitment to the company's values.
- Compensation and Rewards to Support Lean. Compensation to support Lean should be based on worker skills and team performance, while pay plans should be based on collective performance in the context of well-defined and well-understood indicators of quality, cost and delivery.

The third block in describing the main principles, practices, and tools of lean include the elements of the so called "lean culture". The elements of this block are not specific activities but rather general principles. A deep and sustainable implementation of Lean requires that such principles are assumed by the company. Lean culture includes the following principles (Bhasin and Burcher 2006):

- Make decisions at the lowest level to foster participation and take advantage of first-hand information about problems.
- Ensure that there is a strategy of change whereby the organization communicates how the goals will be achieved. People have to be aware of the process of change, its objectives, and its specific steps.
- Develop supplier relationships based on mutual trust and commitment. A real and profound transformation will not be possible unless supplier involvement is guaranteed.

Table 1 Main elements of Lean

Production methods	Work practices	Culture
Just in time Total productive maintenance (TPM) Autonomation Value stream focus Continuous improvement	Standardization, discipline and control Continuing training and learning Team-based organization Participation and empowerment Common values Compensation and rewards to support Lean	Decisions at the lowest level Strategy of change Supplier relationships Focus on the customer Lean leadership Long term commitment

- Systematically and continuously focus on the customer. For the lean company, customer focus is not only a general principle but a critical objective that is served by applying appropriate practices and tools.
- Promote lean leadership at all levels. Leadership style largely determines how work is developed and decisions are taken. Adapting leadership to the lean principles is a critical element of lean implementation.
- Have long term commitment. For a lean company, the implementation of lean practices is not a campaign with limited and specific objectives, but a strategic and long-term policy.

The different elements of Lean that have been mentioned are presented in Table 1. The high number and diversity of practices involved give rise to some confusion regarding what Lean really is, as mentioned before. It must be emphasized that a large number of the practices mentioned are closely related, and can be seen as different ways to define the same functioning. Focus on the Costumer and Value Stream Focus, or Participation and Decisions at the Lowest Level, for example, reflects the same vision and practices from different points of view.

3 SCM and Lean

As discussed in the previous section, Lean is a multifaceted approach, and so is SCM. To analyse the common practices of Lean and SCM, we start from lists of practices defining both approaches. In the case of Lean, an authoritative source is taken. The *Toyota way* principles (Liker 2003) have their origin in the internal documentation of Toyota. They are in consequence the principles of Lean, as seen by the creators of the approach. Production methods, management principles, and more specific practices are presented in Table 2. Aspects directly related to SCM are highlighted in bold. They are principles 2, 3, 4 and 5. Principles 1, 5, 6, 9, 12, 13 and 14 refer to cultural aspects, which are also taken into account by SCM.

The origin of SCM is not clearly tied to a specific company and author, and therefore no equivalent of the 14 Toyota principles for Lean is available. A work

Table 2 Toyota way 14 principles (Liker 2003)

1. Base your management decisions on a long-term philosophy, even at the expense of short-term financial goals 2. Create a **continuous process flow** to bring problems to the surface 3. Use "**pull**" **systems to avoid overproduction** 4. **Level out the workload** 5. Build a culture of stopping to fix problems, to get quality right the first time 6. Standardized tasks and processes are the foundation for continuous improvement and employee empowerment 7. Use visual control so no problems are hidden 8. Use only reliable, thoroughly tested technology that serves your people and processes	9. Grow leaders who thoroughly understand the work, live the philosophy, and teach it to others 10. Develop exceptional people and teams who follow your company's philosophy 11. **Respect your extended network of partners** and suppliers by challenging them and helping them improve 12. Go and see for yourself to thoroughly understand the situation 13. Make decisions slowly by consensus, thoroughly considering all options; implement decisions rapidly 14. Become a learning organization through relentless reflection and continuous improvement

Concepts related to supply chain are highlighted in bold

reflecting the topics dealt with in the literature will be used (Croom et al. 2000). The principle components of SCM literature are shown in Table 3. As the list of topics considered is an extensive enumeration of the literature concerns, the mere presence of a concept does not mean that it is relevant for SCM. Concepts related to Lean are highlighted in bold.

As we can observe, in the blocks "relationships and partnerships" and "best practices" there are many coincidences with Lean. This is not surprising due to the central role of coordination in both approaches and the general applicability of good practices.

A list of principles defining Lean is presented in Table 2, while a list of components of the SCM literature is given in Table 3. The elements of these lists that are common to Lean and SCM are presented in Table 4. They are classified by taking into account three elements present in both approaches: VSM, Continuous Improvement and Cultural Aspects and Relations Management.

From the analysis of the content of Table 4, we can deduce that VSM is similarly important for both Lean and SCM. Concepts with a strategic relevance are common to the two approaches. Continuous Improvement and Cultural Aspects shared by Lean and SCM seem to be more important for Lean than for SC, as half of the 14 principles are involved in it. Conversely, Relations Management aspects that are common to both approaches are more detailed in SCM. On the whole, flow, Continuous Improvement, Cultural Aspects, and Relations Management play an important role both in Lean and SCM.

In addition, differences between the topics addressed by Lean and SCM can be deduced from the same analysis. Aspects regarding Production Methods and Quality, such as TQM, TPM, and Autonomation, are central to Lean and have little or no presence in SCM. The same happens with Work Practices of Lean, as

Table 3 Principal components of supply chain literature (Croom et al. 2000)

Strategic management	*Logistics*
1. Strategic networks	31. Integration of materials and information flows
2. **Control in the supply chain**	32. **JIT**, MRP, waste removal, VMI
3. **Time-based strategy**	33. Physical distribution
4. Strategic sourcing	34. Cross docking
5. Vertical disintegration	35. Logistics postponement
6. Make or buy decisions	36. Capacity planning
7. Core competencies focus	37. Forecast information management
8. Supply network design	38. Distribution channel management
9. Strategic alliances	39. Planning and control of materials flow
10. Strategic supplier segmentation	*Marketing*
	40. Relationship marketing
11. **World class manufacturing**	41. Internet supply chains
12. Strategic supplier selection	42. Customer service management
13. Global strategy	43. Efficient consumer response
14. Capability development	44. Efficient replenishment
15. Strategic purchasing	45. After sales service
Relationships/partnerships	*Best practices*
16. **Relationships development**	46. **JIT**, MRP, MRP II
17. **Supplier development**	47. **Continuous improvement**
18. Strategic supplier selection	48. Tiered supplier partnerships
19. **Vertical disintegration**	49. Supplier associations
20. Partnership sourcing	50. **Leverage learning network**
21. **Supplier involvement**	51. **Quick response, time compression**
22. Supply/distribution base integration	52. **Process mapping, waste removal**
	53. **Physically Efficient versus market oriented supply chains**
23. Supplier assessment	
24. Guest engineering concept	*Organizational behavior*
25. Design for manufacture	54. Communication
26. Mergers acquisitions, joint ventures	55. Human resources management
	56. Employees' relationships
27. Strategic alliances	57. Organizational structure
28. Contract view, trust, commitment	58. Power in relationships
	59. **Organizational culture**
29. Partnership performances	60. Organizational learning
30. Relationship marketing	61. Technology transfer
	62. Knowledge transfer

Concepts related to Lean are highlighted in bold

described in Sect. 2. Topics related to Logistics and Marketing have an important presence in SCM and are secondary for Lean.

4 Lean Supply Chain (LSC)

When Lean and SCM approaches are addressed, LSC must also be considered. LSC consists in applying the Lean concepts to the whole supply chain. In this case, SC refers to the scope and not to specific production management proposals. It can be

Table 4 Classification of common elements between Lean and SCM

Topic	SCM aspects of lean principles (elements of Table 2)	Lean aspects in the SCM literature (elements of Table 3)
VSM	• Continuous process flow (2) • Use "pull" systems to avoid overproduction (3) • Level out the workload (4)	• Time-based strategy (3) • JIT (32, 46) • Quick response, time compression (51)
Continuous improvement and cultural aspects	• Base your management decisions on a long-term philosophy (1) • Build a culture of stopping to fix problems (5) • Standardized tasks and processes (6) • Grow leaders (9) • Go and see for yourself (12) • Make decisions slowly by consensus (13) • Become a learning organization (14)	• Control in the supply chain (2) • Continuous improvement (47) • Process mapping, waste removal (52) • Market oriented supply chains (53) • Organizational culture (59)
Relations Management	• Respect your extended network of partners (11)	• Relationships development (16) • Supplier development (17) • Supplier involvement (21) • Vertical disintegration (19) • Leverage learning network (50)

said that the concept of SC implies an intention to integrate its elements, or, at least, a will to take all its elements into account. Beyond the generic intention to consider the SC globally, we cannot expect that LSC respects all SCM principles.

Some of the key characteristics of Lean manufacturing principles that apply to supply chain strategies are (Ben Naylor et al. 1999): (1) Use of Market Knowledge; (2) Virtual Corporation, Value Stream and Integrated Supply Chain; (3) Lead Time Compression; (4) Eliminate Muda; and (5) Smooth Demand and Level Scheduling. Note that the use of market information is directly related to pull, because knowing the demand is necessary to produce according to demand. The mentioned principles are mainly focused on Value Stream Integration and Production Tools.

A set of 6 principles has been established to define a value chain as Lean (Vitasek et al. 2005). These principles are summarized as follows:

1. Demand Management Capability. In Lean, production is developed to cope with actual customer demand. This implies the need of having detailed immediate information of the demand and managing the flow to cope with this demand.

Table 5 Comparison of the topics treated by Lean, LSC and SCM

Topic	Lean	LSC	SCM
Continuous improvement and cultural aspects	●	•	•
Value stream management	●	●	●
Logistics			●
Marketing			●
Production methods and quality	●	●	
Relations management	•	●	●
Work practices	●		

● = essential
• = taken into account

2. Waste and Cost Reduction. This principle includes preventing wastes, and reducing inventories and times along the entire flow.
3. Process and Product Standardization. The standardization of processes and products enables continuous flow to the customer.
4. Industry Standards Adoption. Standardization also needs to extend beyond a company's particular supply chain to the industry overall. Industry product standards benefit not only consumers but also companies by reducing the complexity of product variations.
5. Cultural Change Competency. There is one recurring obstacle to successfully apply LSC concepts—resistance from the people who will be asked to embrace and implement the change.
6. Cross-enterprise Collaboration. The final attribute of the lean supply chain is Cross-enterprise Collaboration. Through collaborative practices and processes, supply chain partners must work to maximize the value stream to the customer.

A comparison of the principles and topics treated by Lean and SCM is presented in Sect. 3. The results of this comparison, together with the topics dealt with by LSC, are presented in Table 5. This kind of analysis cannot reflect the particular details of each approach, but gives a global vision of its priorities. The main emphasis of Lean is on Culture, Flow, Production Methods and Work Practices. For LSC, VSM and Productions Methods remain important and so does Relations Management. Although present, continuous Improvement and Cultural Aspects are not so important, probably because the emphasis is on aggregate aspects. Work Practices are not dealt with by LSC. On the other hand, SCM gives importance to Logistics, Marketing and Relations Management, and not to Continuous Improvement and Cultural Aspects, while Production Methods are not among its typical topics. Finally, it is important to remark the central role of VSM in the three approaches. The following section is devoted to it.

5 Value Stream Management

We conclude that VSM is the most important point in common between Lean, LSC, and SCM (Sects. 2 and 4). Value stream focus is consubstantial to SCM and is a clear priority for Lean and LSC implementations. In this section, some details of focusing on flow are given.

To be effective, VSM must consider value stream. Value stream is defined as the sequence of activities that are made from the reception of the customer order to the delivery of the product or service (Womack and Jones 1994). Value stream encompasses the production flow from raw material into the arms of the customer, and the design flow from concept to launch (Rother and Shook 2003). The concept of flow along the value stream is graphically represented in Fig. 1. The flow through different companies and departments of each company has to be considered, following the entire path from customer order to the product or service delivery.

Waste reduction and customer focus are two principles that are widely used in Operations Management. It can seem surprising that companies do activities that can be considered waste and take decisions that are not addressed to cope with customers. We can assume that a job is done when it is thought to be relevant. In fact, the problem is in knowing what is really necessary and addressed to customer needs and what is not. In complex value streams, the perceptions about what is needed and what will generate customer value are, in some cases, wrong (Zokaei and Hines 2007). This sometimes results in not needed activities taking place without taking into account the needs of final clients.

The production system applied has a strong influence on the visibility of the flow. The mass production paradigm has emphasized the importance of operational efficiency by pushing goods in large batches into a stable marketplace. This results in massive disconnection between consumers and providers (Womack and Jones 2005). Lean attempts to combine the principles of craftsmanship with mass

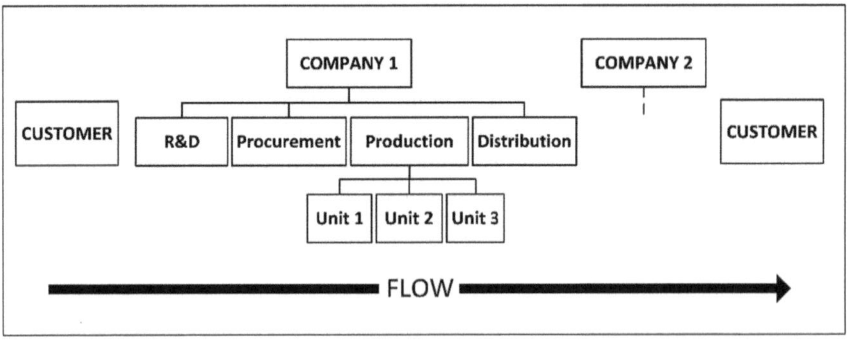

Fig. 1 Graphical representation of the concept of flow

production while avoiding the high cost of the former and the rigidity of the latter (Womack et al. 1990).

Some authors view value stream as a central and strategic concept for Lean transformation. VSM develops the VSM approach into a strategic and holistic method. VSM was defined as a strategic and operational approach designed to help a company or a complete supply chain to achieve a lean status (Hines et al. 1998). The lean enterprise has been considered as a new organizational model that is characterized by the concept of value stream (Womack and Jones 1994). Getting managers to think in terms of value stream is the critical first step to achieving a lean enterprise. It is suggested that someone with real leadership skills and a deep understanding of the product and process must be responsible for the process of creating value for customers and must be accountable to the customers (Liker 2003).

Focusing on value stream requires specific measures to be adopted. A multiple case study analysing what companies do to focus on value stream concluded that they adopt (Olivella and Gregorio 2014):

(1) Organizational units based on value streams, or, when not possible, the appointment of a manager that coordinates the value stream.
(2) Performance measurement system based on value streams.
(3) Formal meeting system focused on value stream issues.

Adaptation of company metrics to value stream has been suggested. It is about eliminating the old metrics and to measure a variety of value stream metrics from lead time to inventory levels to first-pass quality (Liker 2003). To do this, value stream costing has been defined as the process of assigning the actual expenses of an enterprise to value streams, rather than to products, services, or departments (Stenzel 2008). Companies using lean accounting have better information for decision making, have simple and timely reports that are clearly understood by everyone in the company, understand the true financial impact of lean changes, and focus the business on the value created for the customers (Maskell and Kennedy 2007).

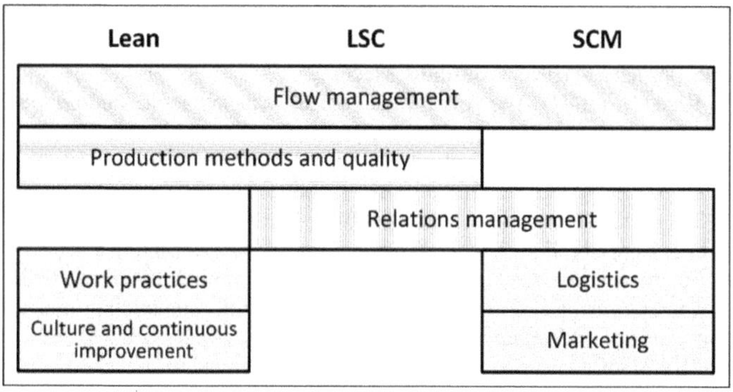

Fig. 2 Graphical comparison of the topics treated by Lean, LSC and SCM

6 Conclusions

In this chapter, the relation between Lean and SCM has been addressed. Both approaches include long lists of principles, practices, and tools. In addition, what Lean and SCM really are is a very controversial topic and very different proposals can be found in the literature. In this context, conclusive results regarding Lean and SCM relations do not seem attainable. The analysis performed tries to reflect which are the topics commonly dealt with by both, Lean and SCM, and those only usual in one of the approaches. Besides, LSC, the application of Lean to SCM, has also been considered as an extension of Lean closely related to the SC concept.

The results are summarized in Table 5 and graphically represented in Fig. 2. Culture and Continuous Improvement, together with Work Practices, are mostly specific to Lean; Productions Methods are shared between Lean and LSC, which has in common with SCM the topics on Relations Management. Logistic and Marketing are specific to SCM. On the other hand, VSM is important for the three approaches. For this reason, details about this concept are in Sect. 5.

References

Belekoukias, I., Garza-Reyes, J. A., & Kumar, V. (2014). The impact of lean methods and tools on the operational performance of manufacturing organisations. *International Journal of Production Research, 52*, 5346–5366. doi:10.1080/00207543.2014.903348

Ben Naylor, J., Naim, M. M., & Berry, D. (1999). Leagility: Integrating the lean and agile manufacturing paradigms in the total supply chain. *International Journal of Production Economics, 62*, 107–118. doi:10.1016/S0925-5273(98)00223-0

Bhasin, S., & Burcher, P. (2006). Lean viewed as a philosophy. *Journal of Manufacturing Technology Management, 17*, 56–72. doi:10.1108/17410380610639506

Corominas, A. (2013). Supply chains: what they are and the new problems they raise. *International Journal of Production Research, 51*, 6828–6835. doi:10.1080/00207543.2013.852700

Croom, S., Romano, P., & Giannakis, M. (2000). Supply chain management: an analytical framework for critical literature review. *European Journal of Purchasing & Supply Management, 6*, 67–83. doi:10.1016/S0969-7012(99)00030-1

Gass, S. I., & Fu, M. C. (2013). *Encyclopedia of operations research and management science*. New York: Springer Science + Business Media.

Hines, P., & Rich, N. (1997). The seven value stream mapping tools. *International Journal of Operations & Production Management, 17*, 46–64.

Hines, P., Rich, N., Bicheno, J., et al. (1998). VSM. *The International Journal of Logistics Management, 9*, 25–42. doi:10.1108/09574099810805726

Liker, J. K. (2003). The Toyota way. McGraw-Hill.

Maskell, B. H., & Kennedy, F. A. (2007). Why do we need lean accounting and how does it work? *Journal of Corporate Accounting & Finance, 18*, 59–73. doi:10.1002/jcaf.20293

Olivella, J., Cuatrecasas, L., & Gavilan, N. (2008). Work organisation practices for lean production. *Journal of Manufacturing Technology Management, 19*, 798–811. doi:10.1108/17410380810898750

Olivella, J., & Gregorio, R. (2014). Organizational practices lean enterprises adopt to focus on value streams. *Journal of Enterprise Transformation, 4*, 309–328. doi:10.1080/19488289.2014.919973

Power, D. (2005). Supply chain management integration and implementation: A literature review. *Supply Chain Management: An International Journal, 10*, 252–263. doi:10.1108/13598540510612721

Rother, M., & Shook, J. (2003). Learning to see: Value stream mapping to add value and eliminate Muda. Lean Enterprise Institute.

Shah, R. (2003). Lean manufacturing: Context, practice bundles, and performance. *Journal of Operations Management, 21*, 129–149. doi:10.1016/S0272-6963(02)00108-0

Spear, S., & Bowen, H. K. (1999). Decoding the DNA of the Toyota production system. *Harvard Business Review, 77*, 96–108.

Stenzel, J. (2008). Lean accounting: Best practices for sustainable integration. Wiley.

Swamidass, P. M. (2000). Encyclopedia of production and manufacturing management. Kluwer Academic Publishers.

Vitasek, B. K., Manrodt, K. B., & Abbott, J. (2005). What makes a lean supply chain? *Supply Chain Management Review, 8*, 39–45.

Womack, J., & Jones, D. (1994). From lean production to lean enterprise. *Harvard Business Review, 72*, 93–103.

Womack, J. P., & Jones, D. T. (2005). Lean consumption. *Harvard Business Review, 83*, 58–68.

Womack, J. P., Jones, D. T., & Roos, D. (1990). The machine that changed the world. Simon and Schuster.

Zokaei, K., & Hines, P. (2007). Achieving consumer focus in supply chains. *International Journal of Physical Distribution & Logistics Management, 37*, 223–247. doi:10.1108/09600030710742434

Multi-echelon Supply Chain Optimization: Methods and Application Examples

Marco Laumanns and Stefan Woerner

Abstract Optimization and optimal control of multi-echelon supply chain operations is difficult due to the interdependencies across the stages and the various stock nodes of inventory networks. While network inventory control problems can be formulated as Markov decision processes, the resulting models can usually not be solved numerically due to the high dimension of the state space for instances of realistic size. In this paper the application of a recently developed approximation technique based on piece-wise linear convex approximations of the underlying value function is discussed for two well-known examples from supply chain optimization: multiple sourcing and dynamic inventory allocations. The examples show that the new technique can lead to policies with lower costs than the best currently known heuristics and at the same time yields further insights into the problem such as lower bounds for the achievable cost and an estimation of the value function.

Keywords Supply chain optimization · Markov decision processes · Approximate dynamic programming · Value iteration

1 Introduction

Supply Chain Management, in particular production and inventory management, is a classical scientific research area in Operations Research. The desire to organize the resource usage and material flow through the value chain in the most efficient way has inspired the development of suitable mathematical models and analytic as well as computational solution techniques for over a century. Many cornerstones of inventory theory are now an integral part of every Operations Research textbook, providing essential insights into the fundamental trade-offs between resource cost and service level and how to optimally parameterize and control inventory systems, even under uncertainty.

M. Laumanns (✉) · S. Woerner
IBM Research, Saeumerstrasse 4, 8803 Rueschlikon, Switzerland
e-mail: mlm@zurich.ibm.com

A.P. Barbosa-Póvoa et al. (eds.), *Optimization and Decision Support Systems for Supply Chains*, Lecture Notes in Logistics,
DOI 10.1007/978-3-319-42421-7_9

Yet, many traditional approaches and tools cannot readily cope with the growing complexity and size of today's supply and production networks, and results that hold for simple systems (such as single warehouse or serial systems) do not extend to general networks. In fact, except for a few special cases, inventory network control problems are usually intractable, and a common approach is to resort to simple heuristics that are known to work well in simpler situations.

Another way to deal with complex problems is by approximation, that is, to solve a problem approximately with less effort, but ideally with some performance guarantees. This paper explores the question whether a recently developed type of approximate dynamic programming, based on piece-wise linear convex value function approximations, is useful approximation technique for typical supply chain control problems. The next section motivates and describes the technique, which is then applied to the well-known problems of multiple sourcing (Sect. 3) and dynamic inventory allocation (Sect. 4), respectively.

1.1 Monotone Approximate Relative Value Iteration

Using appropriate definitions of state and action spaces, multi-echelon supply chain control problems can conveniently be formulated as Markov decision processes (MDPs). However, as solving an MDP explicitly requires to store values for all states, the curse of dimensionality usually makes it impossible to solve such models for supply chain control problems of realistic size.

One way to cope with the curse of dimensionality is to look for an approximate solution for the MDP, for instance using Approximate Dynamic Programming (ADP). ADP is an active research area with a variety of literature. Standard text-books include, e.g., Bertsekas and Tsitsiklis (1996) or Powell (2007).

A typical approach in ADP is to approximate the optimal value function using a linear combination of a fixed set of functions on the state space, called basis functions. Such approaches were shown to work well for some applications (see e.g. Schweitzer and Seidmann 1985 or Farias and Van Roy et al. 2006). However, determining a good set of basis functions is often very difficult and very problem-specific.

In this section, we propose a generic approach for supply chains by exploiting the fact that many supply chain control problems can be described as controlled linear systems in discrete time with the following features:

- The state space is given by a polyhedral set $S := \{x \in R^n | Qx \leq \beta\}$.
- The set of feasible actions U(x) for a given state $x \in S$ is defined as

  $U(x) := \{u \in R^m | Ru + Vx \leq \delta\}$, that is, a polytope for each state x.
- The dynamics are linear so that for any state-action pair (x, u), the random successor state x is given by $x' := Ax + Bu + C$ where (A, B, C) is a multidimensional random variable that takes values in $R^{n \times n} \times R^{n \times m} \times R^n$. We assume

there are $J \in N$ different possible realizations for (A, B, C), denoted by (A_j, B_j, C_j) with probabilities $p_j \in [0, 1]$.

- The immediate costs r: $S \times U \rightarrow R$ are assumed to be piecewise linear and convex.

As shown in Woerner et al. (2015), there exists a convex solution for the discounted cost as well as for the average cost optimality equation for SLCPs. This motivates the use of piece-wise linear convex approximations of the value function (instead of a linear combination of basis functions as in usual ADP approaches). Therefore we use a specialized algorithm called MARVI (Monotone Approximate Relative Value Iteration, see Woerner et al. 2015), to obtain a piece-wise linear convex value function approximation for it.

Any such value function approximation h: $S \rightarrow R$ defines a corresponding policy as

$$\pi_h(x) - \arg\min_{u \in U(x)} \left\{ r(x,u) + \sum_{j=1}^{J} p_j h\left(A_j x + B_j u + C_j\right) \right\}$$

Thus, instead of fixing a set of basis functions as in standard ADP approaches, we take the maximum of a set of hyperplanes, which will be adaptively generated during the run of the algorithm.

2 Application for Multiple Sourcing

The multiple sourcing problem is a well-studied problem in inventory control (Minner 2003). The goal is to optimally control a single stock by continuously ordering from a given set of potential suppliers. In this setup, each supplier represents a different combination of ordering cost and lead time.

Figure 1 gives a schematic view of the multiple sourcing problem for the case of two suppliers. Supplier S_1 has a longer delivery lead-time, in this case three time steps, which is indicated by its order queue the ordered items have to pass until they reach the inventory node. In contrast, the orders of the faster supplier S_2 are available already in the next time step after an order is placed. Using the notation of the previous section, x represents the state (the inventory level and the open orders) and u represents the action (the ordering decisions at the two suppliers).

The multiple sourcing problem is a well-studied hard problem. The goal is to find an optimal policy (a mapping from states to actions) that minimizes the infinite horizon average cost, which consists of ordering cost, inventory holding cost and backlog cost for unmet demand.

Since nothing is known about an optimal policy, different heuristics have been developed and are commonly used in practice:

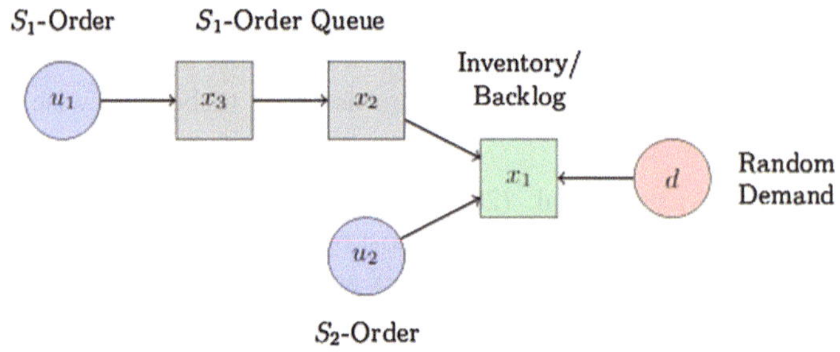

Fig. 1 Schematic view of the multiple sourcing problem for two suppliers

- The Fixed Order Ratio Policy simply allocates a fixed ratio of the order quantity to each supplier in each time step.
- The Constant Order Policy orders a constant amount at one supplier (usually the slower one) and reacts to variability via optimally ordering from the other supplier. The problem is thereby reduced to a standard one-warehouse inventory control problem, which can be solved efficiently.
- The Single Index Policy uses the inventory position (inventory level plus unfilled orders) as an index to be compared to a threshold to determine how much to order at each of the suppliers.
- The Dual Index Policy uses the inventory position as well as the expedited inventory position (inventory level plus unfilled orders that will arrive within the lead time of the fast supplier) to determine how much to order at the slow and the fast supplier, respectively.

The Dual Index Policy has been shown to perform very well compared to the optimal policy on a set of small instances where the optimal policy could be computed by numerically solving the corresponding MDP (Veeraraghavan and Scheller-Wolf 2008).

We now compare the Dual Index Policy with the policy found by applying the MARVI algorithm as described in the previous section. The following problem parameters are chosen:

- Holding cost: 1$ per time step,
- Backlog cost: 5$ per time step,
- Ordering cost: 1$ per unit for the slow supplier S_1 and 2$ per unit for the fast supplier S_2.

The delivery lead time of the fast supplier is fixed at zero, while we consider different problem instances where the lead time of the slow supplier varies from 1 to 6 time steps.

Figure 2 displays the resulting total costs as a function of the lead-time of the slow supplier, for both the optimal Dual-Index Policy and a MARVI policy. The results show that the MARVI policy obtains lower total costs than the Dual Index Policy, and that this gap tends to increase with increasing lead-time difference

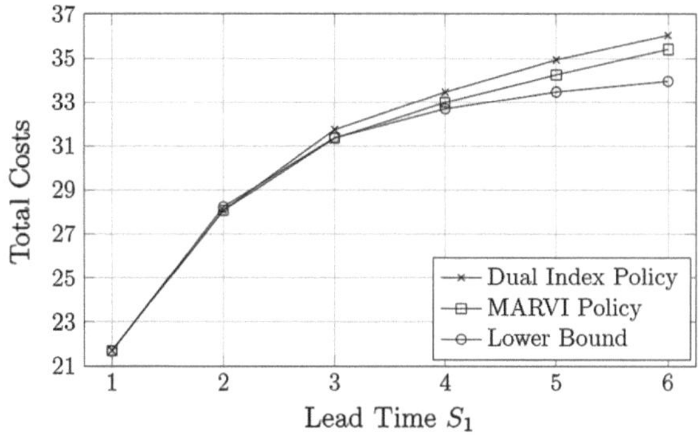

Fig. 2 Total costs for the dual sourcing problem as a function of the lead-time of the slow supplier

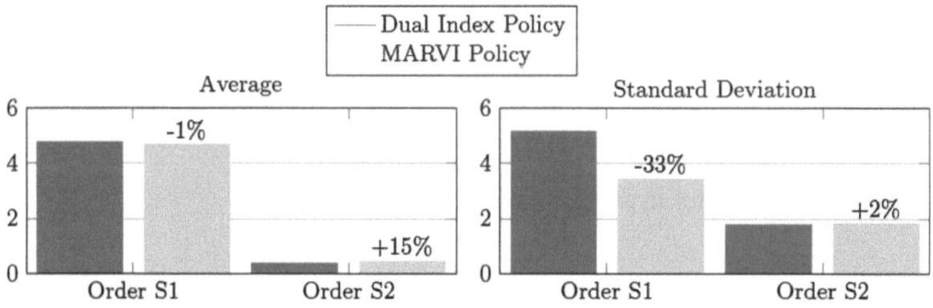

Fig. 3 Average order quantities and standard deviation (per time step) of the Dual Index Policy and the MARVI policy

between the suppliers. Furthermore, a lower bound is shown, which is also computed by the MARVI algorithm as a by-product and allows us to give an instance-specific performance guarantee.

Another advantage of the MARVI policy is that it reduces the order variance for the suppliers, as shown in Fig. 3.

3 Application for Dynamic Inventory Allocation

Dynamic inventory allocation is the process of allocating inventory (from a potentially large set of alternative stocking locations) to incoming orders. This task is typically faced in operations of large, spatially distributed inventory networks, such as spare part networks (Tiemessen et al. 2013) or in order fulfillment of large e-commerce retailers.

The key decision in dynamic inventory allocation is therefore the decision from which stocking point a given order should be satisfied. This decision should optimally balance the (immediate) fulfillment cost (which mainly consists of transportation and handling costs) with the expected future opportunity cost of decreasing the inventory at the stocking point chosen for fulfillment. This

Score Card			
Demand	Bronze Customer 1 × Product XYZ		
Delivery Option	Delivery Costs	Exp. Future Costs	Total Costs
Warehouse 1	100$	10$	110$
Warehouse 2	50$	100$	150$
Central Hub	1000$	0$	1000$

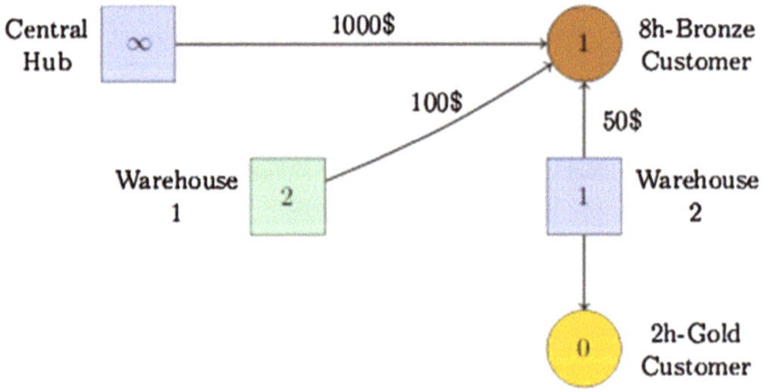

Fig. 4 Decision situation for dynamic inventory allocation

opportunity cost is mainly related to the loss of serviceability in case of future orders (for example lost sales in an e-commerce environment or contract penalties for spare part operations).

The decision situation faced by the dispatcher is schematically represented in Fig. 4 for the example of a spare part request of a customer with contractually guaranteed delivery time of 8 h. The score card lists the different delivery options for the current order, together with the respective delivery costs, and estimation of the expected future opportunity costs and the resulting total cost. The graph below depicts the decision situation geographically.

Besides the customer of the current order and the different delivery options, Fig. 4 also indicates the location of another potential customer. In case the first customer order is satisfied from Warehouse 1, a potential future demand by the second customer might lead to a stock-out situation at Warehouse 1, and consequently, if no other delivery options are available, to a violation of the service level agreement and hence a penalty payment. This risk is captured in the expected future costs.

The main question is how to compute the expected future cost. If this value was known, the optimal policy would be to allocate the order to the delivery option with lowest total costs. As in the previous sections, the problem can be formulated as a Markov decision process with the inventory levels at the different warehouses, the open replenishment orders (due to the replenishment lead-times), and the open customer demand as the state variables and the allocation as well as replenishment decisions as control actions. Figure 5 shows such a network with a central hub

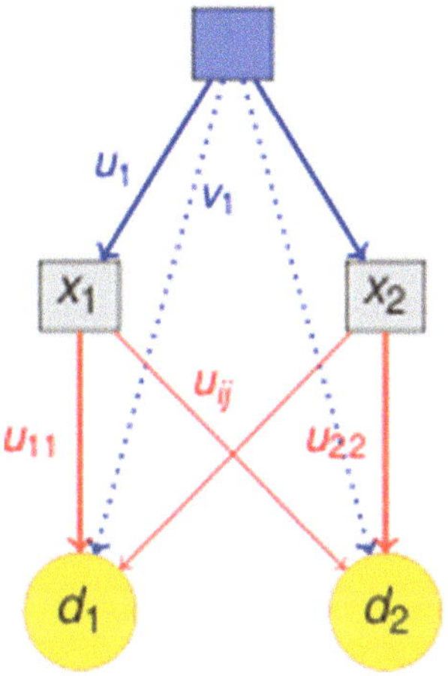

Fig. 5 Simplified state space representation of the dynamic inventory allocation problem

(top), two stocking locations (center), and two customers (bottom), as well as all possible delivery options.

Again, solving the MDP model for an inventory allocation network of realistic size is prohibitive and thus heuristic or approximate techniques are needed.

In a case study with approximately 40 warehouses and 150 customers the MARVI algorithm was applied to approximate optimal inventory allocation policies. Policies were computed separately for 18 different products. Compared to a greedy heuristic that simply allocates based on lowest immediate delivery cost, the study has shown that, depending on the instance, a total cost reduction of 1 % to 10 % was achievable. Besides optimizing the fulfillment decisions, a further advantage of applying MARVI is that it also yields an estimation of the expected future cost, which is very useful information for the dispatcher, as it gives a clear quantitative rationale why the suggested fulfillment option is optimal.

4 Conclusions

Two typical problems in multi-echelon supply chain management have been discussed in this paper, one regarding the supply side (multiple sourcing) and another one regarding the demand side (dynamic inventory allocation) of the supply network. Both problems are hard, so that a direct solution by solving the corresponding Markov decision process is usually impossible due to the size of the state space. Nevertheless, both problems are widely encountered in practice and need to be dealt with somehow, usually by simple heuristics.

This paper has discussed the application of a new approximation technique—Monotonous Approximate Relative Value Iteration—that exploits the structural properties that these problems (and many other supply chain control problem) have, specifically their linear dynamics and piece-wise linear convex cost structure. The main advantages of this approach are: (i) it handles high-dimensional state spaces, (ii) it provides a policy on any approximation it builds (even on a coarse approximation) and (iii) it provides lower bounds. The main disadvantages are: (i) the approach is algorithmically more difficult to implement than a standard dynamic programming approach and (ii) the policy as well as the bounds might be weak if the approximation is not good enough. A limitation is given by the assumption of piece-wise linear and convex costs, so the approach is not able to handle fixed costs.

Whether the mentioned disadvantages and limitations prevent the approach to yield meaningful results is case specific. On the examples considered in this chapter, the approximation technique achieves a noticeable (up to 10 %) cost reduction compared to the best-know heuristics while at the same time giving valid lower bounds on the total cost. In future work, the applicability of the technique to other problems in supply chain management should be considered and its scalability to larger state space dimensions should be explored. One goal of further investigation would be to analyze whether certain problem properties, such as the mixing properties of the underlying Markov chain or the number of different realizations of the random variables, are indicative of the complexity of the optimal value function, its approximability, and thus the quality and performance of our approximation approach.

References

Bertsekas, D. P., & Tsitsiklis, J. N. (1996). *Neuro-dynamic programming*. Athena Scientific.

Farias, V. F., & Van Roy, B. (2006). Tetris: A study of randomized constraint sampling. In *Probabilistic and randomized methods for design under uncertainty* (pp. 189–201).

Minner, S. (2003). Multiple-supplier inventory models in supply chain management: A review. *International Journal of Production Economics, 81–82*, 265–279.

Powell, W. B. (2007). *Approximate dynamic programming: Solving the curses of dimensionality*. Wiley-Interscience.

Schweitzer, P. J., & Seidmann, A. (1985). Generalized polynomial approximations in Markovian decisions processes. *Journal of Mathematical Analysis and Applications, 110*(2), 568–582.

Tiemessen, H., Fleischmann, M., van Houtum, G.-J., van Nunen, J., & Pratsini, E. (2013). Dynamic demand fulfillment in spare parts networks with multiple customer classes. *European Journal of Operational Research, 228*(2), 367–380.

Veeraraghavan, S., & Scheller-Wolf, A. (2008). Now or later: A simple policy for effective dual sourcing in capacitated systems. *Operations Research, 56*(4), 850–864.

Woerner, S., Laumanns, M., Zenklusen, R., & Fertis, A. (2015). Approximate dynamic programming for stochastic linear control problems on compact state spaces. *European Journal of Operational Research, 241*(1), 85–98.

Optimization Lab Sessions: Major Features and Applications of IBM CPLEX

Juan Manuel Garcia-López, Kseniia Ilchenko and Olga Nazarenko

Abstract The Optimization and Decision Support Systems for Supply Chains (Odss4SC) summer school has taken place in Portalegre, Portugal, for three editions (2012, 2013 and 2014) within the Erasmus Intensive Programme. The audience is formed by M.Sc./Ph.D. students on Engineering and Logistics, coming from multiple countries. Some of the goals of this school include studying "green" logistics aspects and using optimization models for solving specific problems, with a practical approach supported by computational sessions (Optimization Labs). During these Optimization Lab Sessions the participants were introduced to the IBM ILOG CPLEX Optimization Studio to develop optimization projects. Participants were acquainted with the concepts, architecture, components, processes, and procedures necessary to build optimization models. Topics covered in these sessions included working with Optimization Programming Language (OPL) to write mathematical programming (MP) models, linking to data sources, flow control and performance tuning of CPLEX optimizer. After the Lab sessions, the participants had acquired a general idea about Optimization engines and how to use them for their Supply Chain Optimization problems.

Keywords Optimization · Mathematical programming · Constraint programming · IBM ILOG CPLEX

J.M. Garcia-López (✉)
IBM—ILOG Decision Optimization, C/Santa Hortensia, 26-28, 28002 Madrid, Spain
e-mail: jm.garcia@es.ibm.com

K. Ilchenko · O. Nazarenko
Kyiv Polytechnic Institute, National Technical University of Ukraine, 37 Peremohy Ave., Kyiv, Ukraine
e-mail: ilchenko@wdc.org.ua

O. Nazarenko
e-mail: olga.nazarenko@ukr.net

A.P. Barbosa-Póvoa et al. (eds.), *Optimization and Decision Support Systems for Supply Chains*, Lecture Notes in Logistics,
DOI 10.1007/978-3-319-42421-7_10

1 Introduction

An introductory session about IBM CPLEX package was delivered at each edition of the program. IBM ILOG CPLEX Optimization Studio speeds development and deployment of optimization models, combining leading solver engines with a tightly Integrated Development Environment (IDE) and modelling language. Techniques used include linear, quadratic and mixed integer programming. IBM ILOG CPLEX Optimization Studio ensures reliable development and maintenance using a transparent modelling language and intuitive tools for model testing, profiling, and tuning.

The session for each edition had an approximate duration of 8 h, with a mix of presentations delivered by the instructor and guided exercises ("Labs").

The goal aimed with these CPLEX sessions was to introduce the students to real, first-class software for modeling and solving Supply Chain problems. It was not intended as a training session in the product, as the duration was obviously insufficient for achieving an effective proficiency level.

The students were evaluated after the session with a short quiz. In Sect. 2 is an excerpt of the questions presented to the students, with enriched answers that can be also useful for researchers and practitioners. The students were asked to answer 'True' or 'False', but long answers are provided here as explanations. Notions and concepts are revisited, and the following text can support further computational Optimization developments on IBM CPLEX.

Fourteen questions were asked in the lab session of the first course edition (2012). In subsequent editions of the course (2013, 2014) twenty questions were asked. The questions are selected and presented here in two separated sets: one dedicated to topics in Mathematical Programming and Constraint Programming (Sect. 2); other addressing topics and applications of IBM CPLEX (Sect. 3).

2 Topics in Mathematical Programming and Constraint Programming

This section based on fundamental works that present main approaches and methods in Mathematical Programming and Constraint Programming. In accordance to this, the materials are presented as overall results of theoretical investigations of sources (Bradley et al. 1977; Imamoto and Tang 2008; Rossi et al. 2006; Williams 1999).

1. ***Constraint programming (CP) is a branch of mathematical programming (MP).***

False: CP has its origins in computer science. It uses constructive search algorithms to find feasible solutions. MP has its origins in Operations Research, and uses algorithms such as simplex and branch-and-bound, with a focus on finding optimal solutions.

The word "programming" causes a common misconception with inclusion of CP to MP. CP can be described as a "programming" in the sense of a computer program: the statements invoke procedures, and control is passed from one statement to another. Moreover, MP only declares algorithms without real programming. The word "programming" was implemented to MP by George Dantzig's application of linear programming to logistics in the military sphere.

A CP optimization model has the same structure as a MP model: a set of decision variables, an objective function to maximize or minimize, and a set of constraints. Still CP and MP came from different branches of science.

2. ***A feasible solution can be, but is not guaranteed to be, an optimal solution***.

True: For a solution to be optimal, it is by definition also feasible.

Feasible solution is an element of the feasible region. The feasible region is the set of all possible solutions of an optimization problem. Moreover, the feasible solution best matched to the problem is called optimal.

For example, there is the set of marks {1, 2, 3, 4, 5}, where the higher value correlates with better marks. Thus, the optimal (in case, if the student wants to be the best in the class) is 5, but all the marks are feasible. The illustration of feasible and optimal solutions for abstract linear programming problem with maximization of objective function is presented on Fig. 1.

3. ***The following constraint is valid for a linear programming problem, where x and y are variables, z is a data item:*** $2x + 3y \le z^2$.

True: As long as z is a data item, then z^2 is also a data item.

LP concept is based on four main components: decision variables (the values to be determined), objective function (shows how the decision variables affect the cost or value to be optimized), constraints and data. Constraints represent how the decision variables use resources, which are available in limited quantities. Data quantifies the relationships represented in the objective function and the constraints.

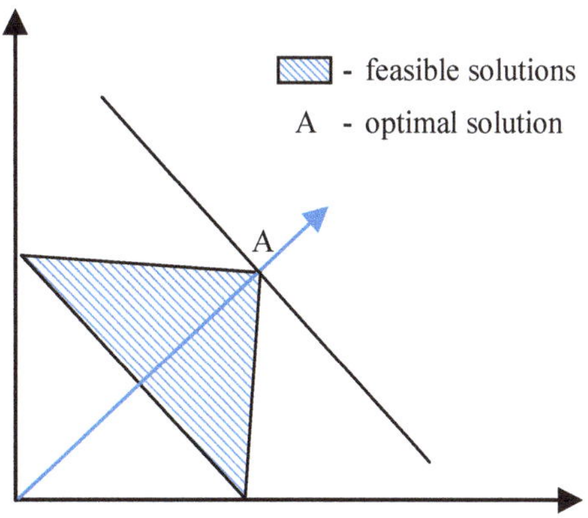

Fig. 1 Feasible and optimal solutions

In a linear program, the objective function and the constraints are linear relationships.

Thus, it is possible to present the CP constraint as $a_1x_1 + a_2x_2 \leq (\geq \text{ or } =)\ z$, where a_1 and a_2—some parameters, x_1, x_2—variables, and z is data.

To sum up, the presented formula is a linear constraint with variables at the left side and data at the right side.

4. ***Hard constraints can be converted to soft constraints to help resolve infeasibilities.***

True: Removing hard constraints and weighting them in the objective function helps in dealing with infeasibilities. However, this technique might yield a problem solution meaningless, if the removed constraint cannot be considered "optional" by the user.

Each inequality that defines the feasible region prohibits certain assignments to the structure. Such constraints are known as hard constraints. If there are many constraints imposed on problem variables, it could be impossible to satisfy them all, such problems are called over-constrained. Therefore, a soft constraint is used for finding some additional decisions. Soft constraints give more scope for relaxing the restrictions and attaching concepts such as cost and priorities to the constraints depending on the preference of the constraint being satisfied in the Constraint Satisfaction Problem. The concept of soft constraints is that constraints are ranked according to their importance and the form that satisfies the maximum high-ranking constraints is considered the optimal.

Hard constraints are those, which we definitely want to be true. These might relate to the successful assembly of a mechanism. Soft constraint are those we would like to be true—but not at the expense of the others. These might say that a mechanism must follow a given path. There is no point in trying to match every point exactly if this can only be done by breaking the assembly of the links.

5. ***An unbounded variable will lead to an objective that can be minimized or maximized to infinity.***

False: An unbounded model will lead to an objective that can be minimized or maximized to infinity. Whether an unbounded model is determined by the constraint definitions.

6. ***A piecewise linear function can be used to approximate convex nonlinear functions.***

True: This technique should not be used for nonconvex nonlinear functions.

A piecewise linear function is a function composed of straight-line sections. A convex function is a continuous function whose value at the midpoint of every interval in its domain does not exceed the arithmetic mean of its values at the ends of the interval. The simplest example of a convex function is an affine function (the sum of a linear form and a constant): $f = a^Tx + b$.

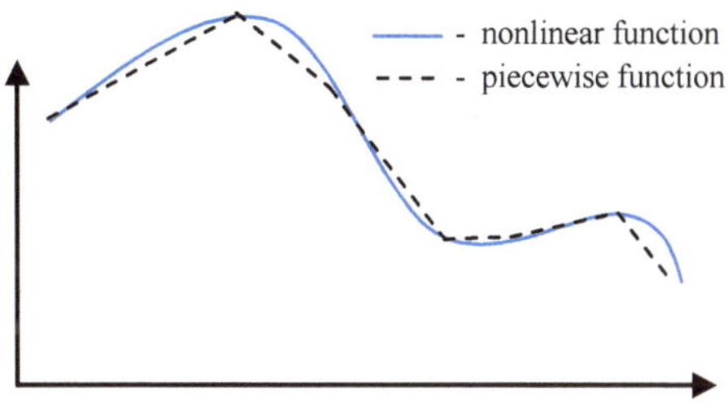

Fig. 2 An approximation of nonlinear function

Approximation of nonlinear function by piecewise linear is a common practice that simplifies the solving. Piecewise functions allow the representation of functions with any accuracy by simply increasing the number of segments until the desired accuracy is met. The approximation of nonlinear function by piecewise linear is shown on Fig. 2.

7. ***Mixed-integer programming is often used for investment planning.***

True: In MIP, for example, the yes/no decisions required for investment planning are represented by binary decision variables. A mixed-integer programming (MIP) problem is such problem where some of the decision variables are constrained to be integer values at the optimal solution.

The typical problem in investment planning can be formulated in such way: an investor needs to plan the budget and the portfolio of the start-up projects for 3 years and has picked up 3 projects he wants to invest in. This problem can be presented in binary way because there are only two decisions: to invest or not to invest in the start-ups. The binary decision variable is 1 if the project is invested in the year, and 0 if there will be no investments for concrete project in concrete year.

Not only investment planning problems are often solved by MIP. The processes of labor planning and logistics are also very popular economic applications for this method.

8. ***Rounding is an efficient method for creating integer solutions to relaxed mixed-integer problems.***

False: Rounding is not the best way to create integer solution to relaxed MIP. Even more, it could give wrong result.

The optimal integer solution can be not obtained by rounding. The closest point to the optimal solution can be even not feasible. Also, the nearest feasible integer point can be far away from the optimal integer point (Fig. 3). Thus, it is not sufficient to simply round the solutions.

9. ***It is important always to use integer variables when a model involves the production of whole items.***

False: While it might be important to use integer variables to plan production of high-value whole items, such as airplanes, it is usually better to use continuous

Fig. 3 Rounding solution in relaxed MIP

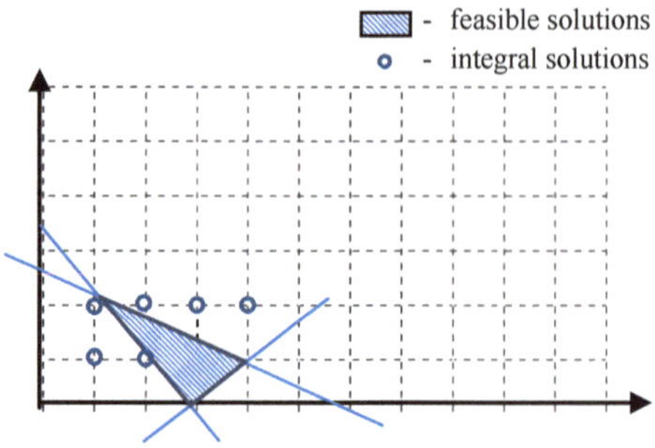

variables to plan for a large number of small items, for example the production of tennis balls, but, in the end, rounding can pose some (small) difficulties.

10. ***Introducing logical constraints in an LP model results in a Mixed-Integer model.***

True. MIP is an optimization method that combines continuous and discrete variables. Involving the logical constraints in an LP model means the usage of binary constraints. Such type of constraints is integer. Therefore, it is possible to speak about a transition to MIP.

11. ***Constraint Programming involves a fine-grained enumeration of time in order to determine start and end times of tasks to be scheduled.***

True. Constraint programming is a type of programming where the relations between variables are described by constraints which must be satisfied at the same time.

A fine-grained description of a system is a detailed model of it. At the same way, the fine-grained enumeration of time helps to determine the start and the end of the process.

12. ***In order to solve a CP problem, you must specify a search phase.***

False: Specifying a search phase is optional, but default search strategy works well in many occasions.

It is necessary to mention that a search phase is a way to guide search types in CP. A search phase defines instantiation strategies to help search the algorithm. A search phase also determines mono-criterion or multi-criteria type of problem. The first one consists of an array of integers to instantiate (or fix), and a variable chooser that defines how the next variable to instantiate is chosen, and a value chooser that defines how values are chosen when variables are instantiated. The second one can have either two different search phases or several decision variables.

CPLEX Studio gives the possibility to change the search algorithms from the IDE settings editor or by changing a CP parameter. In case of missing the specification of search phase, the default algorithm will work. Therefore, the specification of a search phase is desirable but not required.

3 Topics in IBM CPLEX

This section has more practical features, in comparison to the previous one; thus, the answers mainly based on IBM Tutorials.[1,2,3,4,5,6]

13. ***Once an OPL project has been created in CPLEX Studio, it can be solved only from inside CPLEX Studio.***

False: The models created by CPLEX Optimization Studio can be integrated into external applications written in Java, C++, .NET or any other language that is call-compatible with them, with the help of application programming interfaces (APIs). Thus, there is no necessity to rewrite the model in other programming languages. This capability enables to include optimization components to other applications.

14. ***When solving an LP model in CPLEX Studio, the user must choose which LP optimizer (for example, Simplex, Dual-Simplex, or Barrier) to use.***

False: The user is allowed to choose the LP optimizer, but by default CPLEX Optimizer will automatically choose a particular LP optimizer based on the problem structure.

Default settings will result in a call to an optimizer that is appropriate to the class of problem you are solving. However, it is possible to choose a different optimizer for special purposes. An LP problem can be solved with usage of CPLEX optimizers: dual simplex, primal simplex, barrier, and even the network optimizer (if the problem can be described through network structure). The variants of optimizers for each type of problem are presented in Table 1.

---

[1]IBM ILOG CPLEX Optimization Studio OPL Language User's Manual. © Copyright IBM Corp. 1987, 2015. http://www-01.ibm.com/support/knowledgecenter/SSSA5P_12.6.2/ilog.odms.studio.help/pdf/opl_languser.pdf?lang=en.

[2]IBM ILOG CPLEX Optimization Studio CPLEX User's Manual. © Copyright IBM Corporation 1987, 2015. http://www-01.ibm.com/support/knowledgecenter/SSSA5P_12.6.2/ilog.odms.studio.help/pdf/usrcplex.pdf?lang=en.

[3]IBM ILOG CPLEX Optimization Studio CP Optimizer User's Manual. http://www-01.ibm.com/support/knowledgecenter/SSSA5P_12.6.2/ilog.odms.studio.help/pdf/usrcpoptimizer.pdf?lang=en.

[4]IBM ILOG CPLEX Optimization Studio V12.6.2 documentation. http://www-01.ibm.com/support/knowledgecenter/SSSA5P_12.6.2/ilog.odms.studio.help/Optimization_Studio/topics/COS_home.html.

[5]IBM ILOG CPLEX Optimization Studio. http://www-03.ibm.com/software/products/en/ibmilogcpleoptistud.

[6]IBM Decision Optimization Center V3.8 documentation. http://www-01.ibm.com/support/knowledgecenter/SSQVNT_3.8.0/ilog.odms.ide.odm.enterprise.help/ODME/ODMEhome.html.

Table 1 Comparison the optimizers for different types of problems

Type of a problem/optimizer	Dual	Primal	Mixed integer	Network	Barrier
LP	+	+		+ (for some problems)	+
QP	+	+		+ (for some problems)	+
MIP			+		
Network				+	
QCP					+

Therefore, it is possible to solve a LP problem without choosing the optimizer

15. ***The Network Optimizer will always improve the solution time for a model with a network structure.***

False: When faced with a difficult network model, it is a good idea to try the Network Optimizer, although an improvement is not guaranteed.

The ILOG CPLEX Network Optimizer recognizes a special class of linear programming problems with network structure. It uses network algorithms on that part of the problem to find a solution from which it then constructs an advanced basis for the rest of the problem. Each problem has different structure and level of difficulty, thus, there is a high possibility that simple LP models will be solved faster by other optimizers.

16. ***A good way to deal with uncertainty is to solve and compare several scenarios of the same basic model.***

True: Scenario planning is the development of number of the ways for solving the problem, which deals with uncertainty.

Scenario planning involves the development of scenarios that capture a range of plausible future conditions and particularly appropriate in complex situations where uncertainties about future conditions and the effectiveness of management actions are uncontrollable. While there are key steps in the process, there is no single established methodology for conducting scenario planning, or even discrete types of scenario planning approaches. But, in general, scenario planning is the development of number of the ways for solving the problem.

IBM ILOG ODM Enterprise is a software enterprise platform for developing and deploying solutions based on optimization algorithms, e.g. CPLEX, for decision makers, especially in business. IBM ILOG ODM Enterprise facilitates scenario creation and comparison. Therefore the scenario management and analysis are provided by this software.

17. ***Data sparsity can be exploited to create only the essential variables and constraints, thus reducing memory requirements.***

True: CPLEX Studio allows one to create variables and constraints for only the relevant data. This is helpful when dealing with very large problems, where limited memory might become an issue.

Sparsity is defined as the fraction of zeros in a matrix: a sparse matrix has a large number of zeros compared to non-zeros. An optimization model usually contains data structures that are in effect combinations of other data structures, such a tuple that includes other tuples as elements, or a multi-dimensional array that uses a different set to index each dimension. However, not all possible combinations are valid. Therefore, removing these combinations leads to time and memory saving.

It is possible to use tuple sets to instantiate only valid combinations, create a sparse array by indexing on the tuple set and sparse data structures.

To sparse data structure, it is necessary to do the following steps:

- use tuple sets to initialize only valid combinations:

```
tuple ExampleData
{
string ProjectID;
string AppsID;
};
```

- creation of a set of all valid pairs:

```
(ExampleData) ProjectApps=...
```

- creation of a sparse array through Boolean variable:

```
dvar Boolean i [ProjectApps];
```

18. ***CPLEX Studio and ODM Enterprise are used to develop packaged optimization-based applications.***

False: CPLEX Studio and ODM Enterprise are used to develop custom optimization-based applications.

ODM Enterprise allows the development and deployment of applications that are scaled to meet individual customers' needs, from standalone Individual (desktop) configurations to enterprise configurations.

CPLEX Studio is a core part of ODM Enterprise, an analytical decision support toolkit for rapid development and deployment of optimization models. CPLEX Studio is enabled for creation the application for decision-making support. Therefore, it is also an instrument for developing custom optimization-based applications.

Table 2 MP and CP engines

MP engine	CP engine
Usage of a combination of relaxations that were strengthened by cutting-planes and "branch and bound" approach	Making a decision on variables and values and, after each decision, performs a set of logical inferences to reduce the available options for the remaining variables' domains
Usage of a lower bound proof provided by cuts and linear relaxation	Proving an optimality by showing that no better solution than the current one can be found
Requirement of the model falls in a well-defined mathematical category	Not making assumptions on the mathematical properties of the solution space

19. ***CPLEX Optimizer includes both MP and CP optimization engines.***

True. MP and CP are both critical to solving complex planning and scheduling problems. Therefore, both engines include to the CPLEX Optimizer. Some of their differences that complemented each other are presented in Table 2.

20. ***CPLEX Studio is embedded in ODM Enterprise for rapid prototyping.***

True. CPLEX Studio is at the core of applications developed as the integrated system IBM ILOG ODM Enterprise.

This system maintains optimization technology-based analytical decision support. ODM Enterprise enables rapid prototyping the flexible decision support applications. Applications based on ODM Enterprise allow users to adjust assumptions, operating constraints and goals.

21. ***A run configuration in CPLEX Studio can contain more than one model.***

False: A run configuration in CPLEX Studio can contain only one model.

Run configurations are a way of handling model, data, and settings files within a project. Run configurations allows to combine models, data sets and optimizer settings for execution, providing a convenient method for testing a model's behavior across data instances, or test different models on the same data set, or test different settings on model/data combinations without having to resort to writing OPL Script or coding. Basically, a run configuration is a variation of a given project for execution purposes, so it must specify which precise model OPL needs to solve. Therefore, it combines a single model file and, optionally, one or more data files and one or more settings files within the project.

22. ***A tuple is analogous to a database table.***

False: A tuple is the analogous to a ROW in a database table.

A tuple is an ordered set of values represented by an array. Thus, the set of tuples can create the table (Fig. 4).

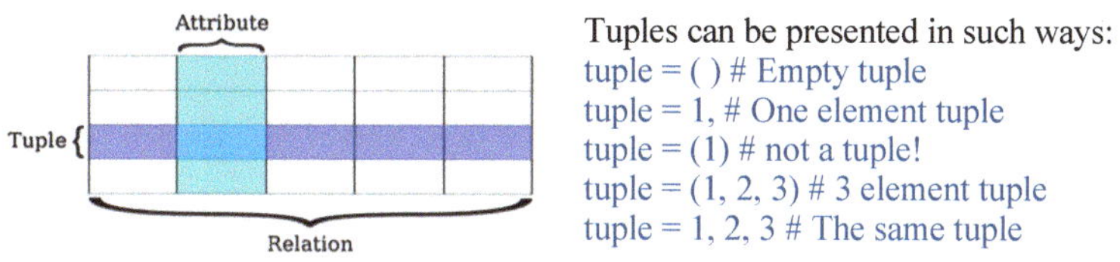

Fig. 4 Tuple presentation in the table form

23. ***The following declaration in the OPL .dat file is correct: float production Cost = 2.5.***

False: The declaration of type 'float' is used in the model, not in the .dat file.

Float is a reserved word for the data type that is used in model files (.mod), but a .dat file is for saving the data in a file, with the extension .dat. Therefore, the declaration is right, but it could not be in .dat file, but in the model file.

24. ***You should label all constraints to be able easily track conflicts.***

False: There is no need to label all of them. However, only labeled constraints are considered by the relaxation and conflict search process.

It is possible to identify constraints by attaching labels to them. This practice is recommended but not required. Furthermore, it is possible to underline both advantages and disadvantages.

Constraint labels enable to benefit from the expand feature in the IDE Problem Browser to find which constraints are tight in a given application or to find dual variable values in linear programs. In addition, it is possible to access the slack and dual values for labeled constraints when a solution is available. Only labeled constraints are considered by the relaxation and conflict search process in infeasible models. However, labels take some performance and memory that can be significant for large models. Nevertheless, there still no need to label all constraints.

4 Conclusions

CPLEX lab sessions were included in the course as a means for the participants to have a close experience with real-world Operations Research software, which they might apply in their future Supply Chain engagements. These lab sessions were short on purpose, to adapt to the tight course schedule. However, all capabilities of the software were presented, to serve as an introductory session rather than as a learning course. The student willing to deepen in his or her skills might then know what to expect and what materials to use.

A quiz with true/false questions was presented to the course participants with the intention to assess their understanding of the topics related to the CPLEX tools. Although some of the questions might be seen as tricky for a "true/false" quiz, this format was chosen for brevity and ease of delivery.

References

Bradley, S., Hax, A., & Magnanti, T. (1977). *Applied mathematical programming*. Reading, MA, USA: Addison-Wesley.
Imamoto, A., Tang, B. (2008). Optimal piecewise linear approximation of convex functions. In *Proceedings of the World Congress on Engineering and Computer Science, WCECS 2008*, October 22–24, 2008, San Francisco, USA.
Rossi, F., van Beek, P., & Walsh, T. (2006). *Handbook of constraint programming* (978 p). Elsevier.
Williams, H. P. (1999). *Model building in mathematical programming*. New York: Wiley.

An Introduction to the Resource Constrained Project Scheduling Problem Solving Techniques

Amaia Lusa and João Luís de Miranda

Abstract A project can be defined as a set of activities. Each activity needs a certain amount of time (that can be variable or even stochastic) and requires different type and quantity of resources (tools, personnel, money, etc.). Also, usually there are additional constraints that affect the scheduling of the activities, such as precedence constraints (for example, an activity cannot start until another one has finished), time constraints (do not start or finish before or later than a certain date) and simultaneously constraints (activities that cannot be done at the same time). In this chapter, different tools for project scheduling will be introduced: graph representation of time and precedence constraints; Gantt chart and load curves; and heuristic and exact algorithms for project scheduling considering different evaluation criteria. Then, a simple case of project scheduling is presented and solved using the described tools, allowing useful comparison between the graph-based approaches and the exact approach of a Mixed Integer Linear Programming (MILP) model implemented in IBM ILOG CPLEX. The implementation and verification procedure followed here allows the enlargement of project applications beyond commercial applications or open literature.

Keywords Project scheduling · Heuristics · Mathematical programming

A. Lusa (✉)
Department of Management, Institute of Industrial and Control Engineering, Universitat Politècnica de Catalunya, Avinguda Diagonal 647, planta 11, 08028 Barcelona, Spain
e-mail: amaia.lusa@upc.edu

J.L. de Miranda
Departamento de Tecnologias e Design, Escola Superior de Tecnologia e Gestão, Instituto Politécnico de Portalegre, 7300-110 Portalegre, Portugal
e-mail: jlmiranda@estgp.pt; joaoluismiranda@tecnico.ulisboa.pt

J.L. de Miranda
CERENA—Centro de Recursos Naturais e Ambiente, Instituto Superior Técnico, University of Lisbon, Ave. Rovisco Pais, IST, 1049-001 Lisbon, Portugal

A.P. Barbosa-Póvoa et al. (eds.), *Optimization and Decision Support Systems for Supply Chains*, Lecture Notes in Logistics,
DOI 10.1007/978-3-319-42421-7_11

1 Introduction

The development of large and complex projects requires specific tools for the planning, scheduling and coordination of a number of activities, tasks, and events, which usually present strong contingence relations among them.

The graph theory and related tools are very useful, namely, the *Program Evaluation and Review Technique* (PERT) whose key aim is to support activities planning and control, while project optimization is not directly addressed. PERT is often used in research projects where significant uncertainty about data arises. For example, to evaluate the probabilities of satisfying completion dates, to identify bottlenecks, to identify the activities requiring more care in order to maintain the program in time, to evaluate project changes and program alterations.

By other side, the *Critical Path Method* (CPM) identifies the sequence of activities that shall be strictly within the schedule otherwise the project terminates in tardiness. CPM allows balancing project costs and time slots, assuming the existence of accurate information about activities, their duration, and the inter-relations between activities and costs, therefore allowing project optimization. CPM is widely applied in projects for civil construction, maintenance programs, etc.

To better illustrate the described tools and discuss their application, a simple case of project scheduling is presented. The case is solved either using graph-related tools that are closely related with PERT and CPM methods, or using a MILP model that is computationally implemented. A useful comparison can thus be developed between these two approaches. This chapter is based on lectures, talks and computational lab sessions held during the triennial period 2012–2014, namely, Lusa (2012, 2013) and Miranda (2013, 2014), and so it is the case of project scheduling under analysis.

The structure of the chapter considers: in Sect. 2, basic notions on project scheduling are presented; in Sect. 3, a simple case of project scheduling is described along with the related solution procedures; in Sect. 4, the MILP model associated with the case at hand is presented and implemented in IBM ILOG CPLEX; in Sect. 5, the main conclusions and comments are presented.

2 Basic Notions on the Project Scheduling Problem

The general problem on project scheduling includes the time specification at which each of the activities in a given time horizon must be started. All activities have known or estimated durations, precedence relations and resource requirements.

The project scheduling objectives usually consider minimizing costs, minimizing all activity finishing times, or even balancing the use of the resources. Examples on real world are common, so as building a plant or a warehouse, developing a new product, devising IT projects, automating a production line, establishing new management software, reorganizing a company, and many others.

To improve the project management and control, scheduling priorities are taking in account, namely: establishing total project duration; estimating the resource requirements and when they will be needed; foreseeing financial and outsourcing needs; redistributing activities to balance resource load and consumption; monitoring and controlling time and money.

The project network is the graph representation of activities and events, including the related precedence relations and time constraints. Different types of project network can be developed accordingly with the edges and nodes attribution, in the current case activities are presented by nodes, while edges represent the relations between activities.

The construction of the project network requires information about activities, it is necessary to define completely each activity and the related attributes. They are commonly accepted: identification through a code and activity description; to know accurately the estimated duration, since the activity may be stochastic, or may depend on available resources; the precedence, time and simultaneously constraints; the required resources, their type and amount.

Some relations or constraints impose conditions on the execution of the all set of activities, and common techniques and schemes are applied, such as:

- Precedence relations between activities, or inter-dependence due to technical factors on different activities are addressed by *precedence constraints*;
- Earliest and latest starting dates of activities, due to commitments, to work force, etc., are treated through *time constraints*;
- Activities that cannot be executed simultaneously, due to a critical resource or technical factors are represented by *simultaneously constraints*;
- Resources availability on staff, technology, materials, money, or others, is emulated using *resource constraints*.

The precedence and time constraints are very important on project scheduling, since they directly define when activities take place. These constraints directly define the earliest and the latest start time for each activity, respectively:

- *Precedence constraints* specify limitations with respect to other activities; for example, on a research project, testing can begin when the prototype is ready (minimum) and final polish must be completed four hours before the end of the thermal treatment (maximum);
- *Time constraints* directly indicate minimum and maximum limitations with respect to a schedule; on an agrarian example, pruning cannot begin until 18 November (minimum) and grape picking must end before 15 October (maximum).

Various significant reasons arise to impose the integration of *simultaneously constraints*. In fact, it is necessary to incorporate information about those activities that need the same critical resource, e.g., activities A and B cannot be executed at the same time. In plus, sometimes activities are performed on the same part of a product or piece, other times activities need the product be put in a different way, e.g., one activity needs the product in one direction and the other in the opposite direction.

The minimum time required to perform all the activities is conditioned by the activities' duration, and by their precedence, time and simultaneously constraints, as well as by resources' availability. The representation of precedence constraints through graphs allows to define the time when all activities can end, the activities that are critical (i.e., those activities whose delay causes overall delays), and also the activities that are non-critical.

2.1 Graph Method

The formalization of precedence and time constraints considers the following parameters:

t_i *i* activity starting date;
t_j *j* activity starting date;
f_i date when activity *i* can begin;
F_i date on which activity *i* must have already started;
a_{ij} minimum time that must elapse between the start of *i* and the start of *j*;
b_{ij} maximum time that can elapse between the start of *i* and the start of *j*;
α the first node of the project network, the starting event;
ω the last node of the project network, the final event.

The minimum time f_i and the maximum time F_i for the start of each activity *i* are defining a feasibility time-window for activity *i* (Fig. 1), but the time that can elapse between the start of activity *i* and the start of activity *j* shall be estimated too. In order to cope with the activities duration, the minimum time and the maximum time between the start of two consecutive activities is also necessary (Fig. 2).

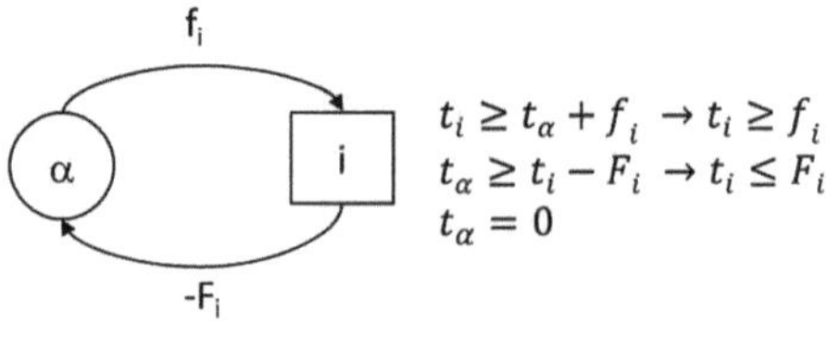

Fig. 1 Minimum (f_i) and/or maximum (F_i) time for the start of activity i

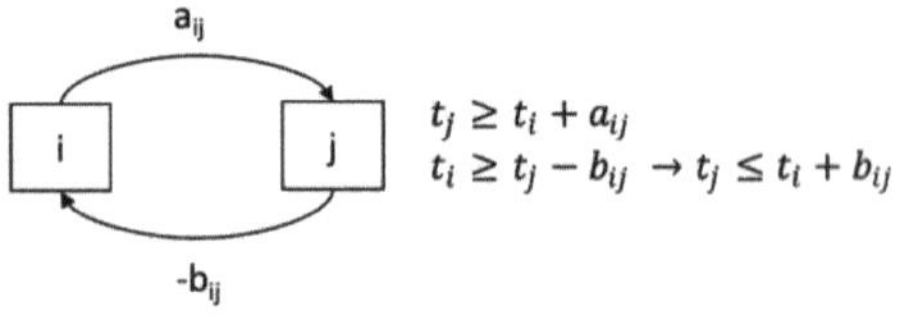

Fig. 2 Minimum time (a_{ij}) and/or maximum time (b_{ij}) between the start of activity i and the start of activity j

From the prior estimates and conjugating with activities duration, the slack time between consecutive activities is calculated and the *critical path* for the project network is identified. The procedure follows the network forward, and then backward, to obtain orderly for each activity *i*:

- *Earliest Start Time* (EST)—An activity's EST is the longest path from α to the activity, and can be calculated with Dijkstra or Ford algorithms. In the case that no return edges nor negative values exist, the Ford algorithm gives the solution in just one iteration (EST is calculated from the first node, moving forward on the network and adding the activities duration time; in other words, EST is calculated by supposing the preceding activities are all done at their earliest time);
- *Latest Start Time* (LST)—the activity's LST is EST(ω) minus the longest path from the activity to ω (the final event, again, can be obtained by using the Ford algorithm); LST is calculated from the last node, moving backward and subtracting the duration time of the preceding activities; this means LST is the latest time for the activity start without provoking any delay in the project time completion.

The activity slack, $Slack_i$, indicates the delay activity *i* can suffer without further implications in the project completion time. For activity *i*, the slack time is defined by the difference between the associated *Latest Start Time* and the *Earliest Start Time*:

$$\mathrm{Slack_i} = \mathrm{LST_i} - \mathrm{EST_i} \tag{1}$$

Therefore, activities with null slack are critical for the project, because any delay will have implications on project completion time. The suggestion is to focus on the control of critical activities, but without forgetting about "sub-critical" activities.

The burden and the time necessary for these calculations, the large number of activities in the project and the related data, the current situation in modeling and solving large optimization problems, and the need to evaluate project schedules on a daily basis are making crucial the utilization of commercial software and formalized methods.

2.2 Heuristic Methods

Problems formulated with integer or binary variables have significant interest but large instances are very difficult to solve, then heuristics are very often applied to obtain admissible and near-optimal solutions. A heuristic is an approximation procedure, aiming at solutions of satisfactory quality that are obtained in a suitable execution time. However, it should be noted that the heuristic approach is to replace the problem exact solution for an alternative solution.

In comparison with exact methods, heuristic methods allow the treatment of very large problems; they rely on simple empiric concepts, as well as heuristics enable the direct use of problem information. However, heuristics do not guarantee solutions optimality and then the inherent deviation on sub-optimal solutions requires quantification.

In general, heuristics are classified accordingly their main purpose:

- Constructive heuristics—the initial and inadmissible solution is conjugated with new elements to construct a possible solution;
- Improvement heuristics—the target is to improve an existing possible solution, e.g., a solution (or group of solutions) that was already obtained by a construction method.

As example cited in Lusa (2012), consider the following pseudo-code for a heuristic method that addresses simultaneous constraints (pairs of activities that cannot be done at the same time and, thus, it must be decided, for each constraint, which activity must precede the other one):

Let N be the number of simultaneously constraints and n the number of already solved constraints;
While n < N do:

- For each constraint and for each direction, calculate a lower bound* for T (time at which all activities have finished)
- Take the constraint-direction with the largest lower bound and solve it in the opposite direction; $n = n + 1$
- Update EST_i and LST_i values

End while
* Direction i → j: Lower bound = "longest path α-i" + di + "longest path j-ω" = ESTi + di + (T-LSTj) (being di the duration of the activity i)

Various types of construction heuristics can be observed:

- Greedy heuristics, for example, when the best option in each iteration is selected, not taking into account the subsequent iterations;
- Random construction, as designated, the solution is set at random; for instance, genetic algorithms consider random alterations as "solutions mutation";
- Construction by simplification, when the problem is separated into different modules or sub-problems, for which are either allowed well-known procedures or the special cases results.

A basic greedy heuristic may consist of a simulation in which activities are scheduled according to some priority rule as, for example the shortest duration or the shortest slack.

As example of improvement heuristics, the Burgess and Killebrew algorithm (1962) allows for distributing load and payments leveling. From the same authors, balance heuristic methods are firstly used to avoid overload, but these methods also level the use of resources. This resources attribute overpasses overload avoidance and the search of feasible solutions: when the starting point is already a solution with no overload on resources allocation, then a better and improved solution can be achieved in the feasible solutions space.

Let Δ be the total overload, $\Delta = \sum_{t=1}^{t_\omega} \max(0, \textit{Required resources} - \textit{Available resources})$

Define;

- Begin with the first activity;
- Select the first activity that can be moved without pushing any other activity, and move it to the spot with the lowest Δ value and as far to the right as possible;
- Go on to the next activity (after the last one, go back to the first one);
- Repeat until $\Delta = 0$ or until it is not possible to move more activities (in the latter case, increase the schedule's total time —t_ω— and start again).

Sometimes heuristics are not producing a feasible solution, therefore the integration of different aspects on solution search can be beneficial. Some examples of heuristic procedures that address precedence and resources constraints follow:

- Use priority rules to build feasible solutions; for example, a greedy algorithm based on building the solution by simulating the execution of the project and selecting the activities to be scheduled (when there are not enough resources to schedule all the activities whose predecessors have finished) according to some priority rule (the shortest activity, the most critical, etc.);
- Balance load by moving non-critical activities, either on an intuitive, by trial and error or using a formalized procedure such as the Burgess and Killebrew algorithm;
- Start out with a solution and improve it by exploring new solutions; the current calculation methods make it possible to explore a large number of solutions in short time, and this type of local optimization methods are quite common nowadays.

In fact, *Local Search* (LS) is a basic tool that starts from a feasible solution, evaluates neighbor feasible solutions with similar structure, and then selects the best neighbor solution according with a specific evaluation function. LS is based on the principle that solutions near optimal solution present also near or similar structures, in despite of being worse solutions. In this method, beyond the definition of the neighborhood structure on search, the evaluation function, the strategy for solution alteration, also the stopping criterion shall be selected. The described LS method allows the introduction of variants, either in the neighborhood search, the alternatives evaluation, or the alterations to be made. The *Iterated Local Search* (ILS) method arises from the repeated iteration of LS method. ILS combines rapid generation of good solutions with asymptotic convergence to the optimum.

In addition, various alternatives to heuristic methods can be selected, in special if the problem dimension is not too large. For example:

- To enumerate and evaluate all the combinations, and then to choose the best one; this alternative can be straightforward if the number of simultaneous constraints is small, but for project scheduling it is not the proper choice;
- To try the *Branch-and-Bound* method (B&B); this is an enumeration method that guarantees optimality, that is, the exact solution is found when the method terminates (for details, see the contribution of Casquilho and Miranda in this book);
- To formulate and solve an optimization MILP model; this option is developed later in Sect. 4, using the same case of Project Scheduling that is described in Sect. 3.

3 A Case of Project Scheduling

A simple case of project scheduling is presented based on Lusa (2012), the related data and the solution procedures are described. The application of basic rules and concepts on project scheduling, the description of some sample calculations, they are both supporting the daily utilization of commercial software and the computational implementation of exact algorithms.

The twelve activities (A–L) of a general project and the associated attributes are given in Table 1; For each activity: the duration in weeks; the preceding activities that must be completed before the activity starts; the people required to perform the activity jobs (renewable resources); and finally, the payments incurred in hundred

Table 1 Project activities and related data: duration, preceding activities, and resources (renewable or not)

Activity	Duration (weeks)	Preceding activities	Renewable resources (people)	Non-renewable resources (payments/week, in h €)
A	1	–	AC&MP	10
B	6	A	RG	5
C	10	B	AC&RG	20
D	18	–	RR	30
E	3	B	AC	10
F	8	E	MP	20
G	1	F	RG	50
H	3	G	BC	20
I	14	B	RR	10
J	3	C	MP	15
K	1	I, J	BC	20
L	1	D, G, K	RR	30

of Euros (non-renewable resources) for each week (while the activity is being executed).

Noting that a person cannot do more than one activity at a time, then the renewable resources (people: AC, MP, RG, BC, RR) shall be strictly assigned to the activities set in non-overlapping mode.

The all set of procedures consider multiple steps, namely, to:

1. Make the precedence's graph. Compute the *Earliest Start Time* (EST), the *Latest Start Time* (LST) and the slack of each activity and determine the critical path.
2. Draw the Gantt and load charts scheduling each activity at its EST.
3. In case the prior schedule is not feasible within the renewable resources (people), determine a feasible schedule by using a heuristic method. Compute the total payment for each period.
4. Modify the schedule found in point 3 with the objective of levelling the use of the non-renewable resources (payments) using the Burgess and Killebrew methods.

Addressing the **first point**, the precedence's graph is constructed by direct observation of the three first columns on Table 1, and specifying the starting event as the network first node, α, at week 0. Thereafter, other nodes follow: activities A and D have no preceding activities; but before starting activity B, firstly activity A shall be completed and it takes 1 week; before starting activity C, activity B shall be completed too and it takes plus 6 weeks, meaning that EST for activity C sums 7 weeks. The same reasoning applies for activities E and I that are only able to start after B completion; and with this on mind the first two columns of Table 2 (the activities and related EST) can be filled in.

The final event or the last node of the project network, ω, occurs at week 23. Activity L has no immediate successors, 1 week duration, and its EST is 22; in plus, activity H also has no immediate successors, the related EST is 19, it takes 3 weeks to be completed, thus activity H is completed at week 22. This way, the

Table 2 Project activities and related EST, LST, and slack times

Activity	EST	LST	Slack
A	0	0	0
B	1	1	0
C	7	8	1
D	0	4	4
E	7	8	1
F	10	11	1
G	18	19	1
H	19	20	1
I	7	7	0
J	17	18	1
K	21	21	0
L	22	22	0

project makespan is 23 weeks, corresponding to EST(ω) because activity L is being completed only at week 23.

The LST column is calculated backwards and starting from the network last node, ω, defining LST(ω) = EST(ω) = 23 and subtracting the duration of preceding activities. L duration is 1 week, then LST(L) is 22, but LST(H) is 20 because its duration is 3 weeks. From Table 1, L preceding activities are K, G, and D; their incumbent LST are LST(K) = 21, LST(G) = 19, and LST(D) = 4 because the related activities duration are, respectively, 1, 3, and 18 weeks. Following the same procedure, the LST values for all set of project activities are indicated in the third column of Table 2.

The slack for each activity can be directly calculated on Table 2 by the difference between LST and EST, the related slack times are indicated in the fourth column. The critical path is thus identified by the activities with zero slack, that is, the critical path considers the activities with equal EST and LST and, consequently, they are not tolerating any delay.

Graph representations are key tools for project scheduling because all the relationships between activities are placed in perspective, namely, activities duration, predecessor nodes for each activity, resources availability, that is, the all set of attributes is shown. Graphs and networks are often used to analyze projects and to support the development of PERT and CPM procedures (PERT puts the activities on the edges, while CPM put them on the nodes). The graph representation for the current project network, aggregating EST and LST values in "EST/LST", and identifying the critical path on dashed line, follows on Fig. 3 (activities are put on the nodes).

Addressing the **second point**, each activity is scheduled at its EST time in the Gantt chart at Fig. 4. One row is allocated to each activity A–L, the starting and the finishing times the activities in the current project schedule are thus directly indicated. Duration times and the preceding relationships are also integrated, all set of activities attributes are taking in account and thus Gantt charts are very popular on practice.

Moreover, Gantt charts can also be used to evaluate resources' load and payments schedule, that is, resources utilization can be fully appreciated. Activities

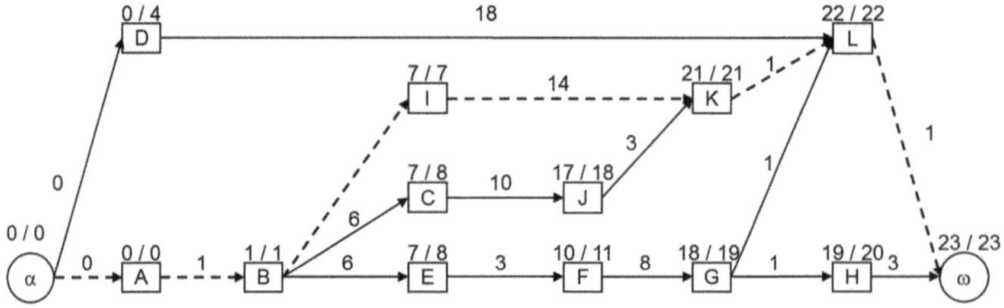

Fig. 3 Graph representation of project network, critical path, EST/LST, and duration times for project activities

		Start time																						
		0	1	2	3	4	5	6	7	8	9	10	11	12	13	14	15	16	17	18	19	20	21	22
		A																						
			B	B	B	B	B	B																
									C	C	C	C	C	C	C	C	C	C						
		D	D	D	D	D	D	D	D	D	D	D	D	D	D	D	D	D	D					
									E	E	E													
												F	F	F	F	F	F	F	F					
																				G				
																					H	H	H	
									I	I	I	I	I	I	I	I	I	I	I	I	I	I		
																			J	J	J			
																							K	
																								L
People	AC	1							2	2	2	1	1	1	1	1	1	1						
	MP	1										1	1	1	1	1	1	1	2	1	1			
	RG		1	1	1	1	1	1	1	1	1	1	1	1	1	1	1	1		1				
	BC																				1	1	2	
	RR	1	1	1	1	1	1	1	2	2	2	2	2	2	2	2	2	2	2	1	1	1		1

Fig. 4 Gantt chart and resources' load for the case project

overlapping, payments leveling, and further improvements on project scheduling can be supported from chart-based analysis, as described in the following text.

Reminding the definition of people as project's renewable resource, one row is allocated to each person (AC, MP, RG, BC, RR) in Fig. 4, and the activities each person is required shall be verified from Table 1. For example:

- Person AC is required in three activities (A, C, and E), and the related starting and finishing times for these activities are directly indicated in the chart; however, 2 activities (C and E) are overlapping weeks 7–9 as shown in the Gantt chart (Fig. 4), then this situation is presented in bold at Fig. 4.
- Person MP is also required in three activities (A, F, and J), the related starting and finishing times are also indicated; by direct observation of the Gantt Chart, an overlapping situation also arises between 2 activities (F and J) at week 17, and bold line is applied once again.
- Other overlapping situations are identified with the same verification procedure, namely, for person BC (week 21, activities H and K) and person RR (weeks 7–17, activities D and I).

Figure 4 allows the reader to identify two activities occurring at same time but each person cannot do more than one activity at a time. People unavailability means (renewable) resources unavailability, the identified activities cannot be performed, and the current project schedule is not feasible with the available resources.

Addressing the **third point** of the current case, a schedule with no overload in the use of the renewable resources (people) shall be established. Usually, a MILP model with the objective of minimizing the makespan is developed, this way avoiding the burden of trial-and-error and repetitive charts-based calculations.

A constructive heuristic aiming at finding a feasible solution can also be developed. The balancing heuristic method presented in Sect. 2.2 (Burgess and Killebrew 1962) is utilized and its reasoning is described in a colloquial manner. Based on a priority rule, the identified activities that need the same critical resource (activities C and E for person AC; activities F and J for person MP; activities H and K for person BC; and finally activities D and I for person RR) are selected; then overload elimination follows a trial-and-error approach that moves activities forward and backwards. For example, initiating the procedure with non-critical activities:

- Activities C and E cannot be executed at the same time; moving EST(C) to week 10 then overload is avoided for person AC; activity C is not in the critical path and it will terminate at week 19; however the alteration of three weeks is over-passing C slack time (1 week), and the project completion time can be delayed;
- Activities F and J cannot be executed at the same time; moving EST(J) to week 18 then overload is also avoided for person MP; since activity J is not in the critical path and this alteration of 1 week is equal to its slack time, then activity J becomes also a critical activity; in plus, activity C is preceding J, then J can start only after C is finished, that is, EST(J) is thus 20;
- Overlapping activities H and K are causing overload for person BC, and this can be avoided by moving EST(K) to week 22; activity K is in the critical path and this alteration will cause 1 week delay on the project completion time;
- However, activities D and I are overloading person RR, the project schedule becomes feasible by moving EST(I) from week 7 to week 18; activity I is also in the critical path, this alteration will cause a large delay (11 weeks) on the project schedule;
- In fact, activity I is preceding activities K and L on the critical path and its duration is 14 weeks; activity I finishes at week 31, then EST(K) is 32 and EST (L) is 33.

The other non-critical activities can be accommodated in this larger time for project completion, 34 weeks, and the related Gantt chart with no overload on people is presented in Fig. 5. This project network corresponds to a feasible solution created through a priority rule that preferably selects non-critical activities, since moving critical activities will cause delay on project termination.

From the no overload project schedule on Fig. 5, and revisiting resources data (renewable and non-renewable) from Table 1, the chart on resources load and payment's schedule is updated and presented in Fig. 6.

Finally addressing the case's **fourth point**, the no overload schedule found in point 3 is modified with the objective of levelling the use of the non-renewable resources (payments). Note the feasible project schedule presented in Figs. 5 and 6 is created based on a non-feasible solution with people overload.

Start time																																	
0	1	2	3	4	5	6	7	8	9	10	11	12	13	14	15	16	17	18	19	20	21	22	23	24	25	26	27	28	29	30	31	32	33
A																																	
	B	B	B	B	B	B																											
										C	C	C	C	C	C	C	C	C	C														
D	D	D	D	D	D	D	D	D	D	D	D	D	D	D	D	D	D																
							E	E	E																								
										F	F	F	F	F	F	F	F																
																						G											
																							H	H	H								
																		I	I	I	I	I	I	I	I	I	I	I	I	I	I		
																				J	J	J											
																																K	
																																	L

Fig. 5 Gantt chart (no overload)

	Start time																																	
	0	1	2	3	4	5	6	7	8	9	10	11	12	13	14	15	16	17	18	19	20	21	22	23	24	25	26	27	28	29	30	31	32	33
AC	1							1	1	1	1	1	1	1	1	1	1	1	1	1														
MP	1										1	1	1	1	1	1	1	1			1	1	1											
RG		1	1	1	1	1	1				1	1	1	1	1	1	1	1	1	1			1											
BC																								1	1	1							1	
RR	1	1	1	1	1	1	1	1	1	1	1	1	1	1	1	1	1	1	1	1	1	1	1	1	1	1	1	1	1	1	1	1		1
PAY	40	35	35	35	35	35	35	40	40	40	70	70	70	70	70	70	70	70	30	30	25	25	75	30	30	30	10	10	10	10	10	10	20	30

Fig. 6 Resources load (no overload)

At now, an improvement heuristic considers balance load by moving the non-critical activities, either on an intuitive way or by trial and error. Even without overload to distribute, the Burgess and Killebrew procedure is useful since it allows payments levelling. Figure 6 shows larger payments during weeks 10–17, due to the simultaneous occurrence of three activities (C, D, and F). The payments evolution can be smoother, on average about 38.67 €, that is, the total payment (1315 €) divided by the makespan (34 weeks). For example, considering as a measure of payments overload the payments above the average value and from the observation of Figs. 5 and 6:

- C and J are non-critical activities, and if they are moved forward it is not expected impact on project completion time; Activity C is preceding the activities sequence J, K, and L; then J activity is moved as far to the right as possible, on weeks 29–31 where the payments are at lower level;
- Thus by the same rationale C can finish at week 28 and start at week 19;
- H activity can also be moved to the spot with the lowest payments and as far to the right as possible, on weeks 29–31; in the same way, this alteration does not affect the project time;

Start time																																	
0	1	2	3	4	5	6	7	8	9	10	11	12	13	14	15	16	17	18	19	20	21	22	23	24	25	26	27	28	29	30	31	32	33
A																																	
	B	B	B	B	B	B																											
																			C	C	C	C	C	C	C	C	C	C					
D	D	D	D	D	D	D	D	D	D	D	D	D	D	D	D	D	D																
							E	E	E																								
										F	F	F	F	F	F	F	F																
																		G															
																													H	H	H		
																		I	I	I	I	I	I	I	I	I	I	I	I	I	I		
																													J	J	J		
																																K	
																																	L

Fig. 7 Gantt chart with payments leveling

- G activity can be performed earlier and after preceding activity F is completed, at week 18 where a lower payment occurs; this alteration avoids that G is performed at same time as C and I, as presented in Fig. 7.

From the Gantt chart with payments leveling on Fig. 7, and revisiting again Table 1 for resources data, the chart on resources load and payment's schedule is updated in Fig. 8.

Comparison between payment schedules follows in Fig. 9: while the initial approach in graph ***a*** presents larger alterations on payments, graph ***b*** shows a smoother evolution of payments along the project time horizon. In fact, the observed alterations on payments at graph ***b*** are of lower amplitude, and the lower variability is very important for finance planning.

Moreover, this solution concerning payments levelling can be also obtained in a two-steps procedure, by solving two MILP models in sequence: (i) Minimizing makespan; and (ii) Minimizing payments irregularities and ensuring the makespan from first step. Other MILP developments for project scheduling problems are treated by Artigues et al. (2013) and Koné et al. (2011).

	Start time																																	
	0	1	2	3	4	5	6	7	8	9	10	11	12	13	14	15	16	17	18	19	20	21	22	23	24	25	26	27	28	29	30	31	32	33
AC	1							1	1	1										1	1	1	1	1	1	1	1	1	1					
MP	1										1	1	1	1	1	1	1	1												1	1	1		
RG		1	1	1	1	1	1												1	1	1	1	1	1	1	1	1	1	1					
BC																														1	1	1	1	
RR	1	1	1	1	1	1	1	1	1	1	1	1	1	1	1	1	1	1	1	1	1	1	1	1	1	1	1	1	1	1	1		1	
PAY	40	35	35	35	35	35	35	40	40	40	50	50	50	50	50	50	50	50	60	30	30	30	30	30	30	30	30	30	30	45	45	45	20	30

Fig. 8 Resources load with payments levelling

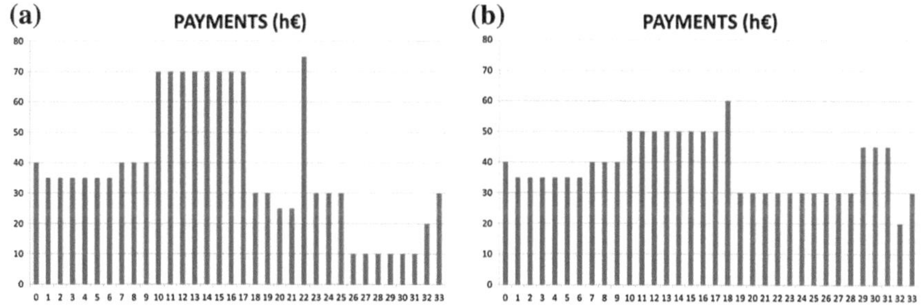

Fig. 9 Comparison between payments schedule: **a** initial approach; **b** levelling the use of the non-renewable resources (payments)

4 Project Scheduling: MILP Model

In this section the implementation of a MILP model that focus the project scheduling case is addressed and we are using the implementation of exact algorithms on commercial software to solve it (Miranda 2013, 2014). The IBM ILOG environment is preferred due to its high performance and friendly utilization, and CPLEX solver is widely selected for Linear Programming (LP), Integer Programming (IP) and Constraint Programming (CP) implementations (IBM 2015a, b, c).

The MILP model is based on Pritsker et al. (1969) formulation, the associated nomenclature such as data, variables, and the model are described in below.

Data:

N Number of activities to schedule

$Tmax$ Upper bound for T (time at which all activities have finished)

d_i Duration of activity i ($i = 1,\ldots,n$)

P_i Set of immediate predecessors of activity i ($i = 1,\ldots,n$)

F Set of activities with no immediate successors

L_i, U_i Lower and upper bound for the starting time of activity i ($i = 1,\ldots,n$)

R Set of types of resources

A_{rt} Quantity of resource r available at time t ($r \in R$, $t = 1,\ldots,Tmax$)

R_{ir} Quantity of resource r necessary to perform activity i ($r \in R$, $i = 1,\ldots,n$)

Variables:

T Time at which all activities have finished

t_i Starting time of activity i ($i = 1,\ldots,n$)

$y_{it} \in \{0,1\}$ Takes value 1 *if* activity i starts at time t ($i = 1,\ldots,n$, $t = L_i,\ldots,U_i$)

The objective function minimizes the project time horizon, that is, the time at which all activities have finished, T:

$$[\min] Z = T \tag{2}$$

The project time horizon, T, shall consider the starting time, t, and the duration, d, of all the activities, i, with no immediate successors:

$$T \geq t_i + d_i, \quad \forall i \in F \tag{3}$$

In a similar way, the starting times of each activity i shall consider the related starting times and the durations of all the activities j that are occurring immediately before activity i:

$$t_i \geq t_j + d_j, \quad i = 1, \ldots, n; \ \forall j \in P_i \tag{4}$$

The logic-binary restriction on starting times t_i is defining one time-value for each activity, i, between the associated lower bound, L_i, and upper bound, U_i, on starting times. The time-value t is an integer number as follows from the sum restriction:

$$t_i = \sum_{t=L_i}^{U_l} t.y_{it}, \quad i = 1, \ldots, n \tag{5}$$

For each activity i, the binary variables y are defined from a special ordered set with one single element equal to one (1), and all the other binary elements are zero (0). This restriction imposes that one and only one starting time is to be selected for activity i:

$$\sum_{t=L_i}^{U_l} y_{it} = 1, \quad i = 1, \ldots, n \tag{6}$$

The model ***Project Scheduling*** is thus conjugating relations 2–6, it aims at the project time horizon minimization and simultaneously considers the restriction on total time, the restrictions set on starting times for all activities, the specification of one integer time-value for each activity, and the associated binary variables specification.

Model *Project Scheduling*:

$$[\min] Z = T \tag{7a}$$

subject to

$$\text{Relations set } 3-6$$

$$t_i \geq 0, \quad \forall i \tag{7b}$$

$$y_{it} \in \{0, 1\}, \quad \forall i, t \tag{7c}$$

The model can be improved by including other scheduling criteria and/or constraints (risk, uncertainty…) and also integrating other decisions (e.g., payments levelling). Some of these advanced topics are discussed in other parts of this handbook, namely, by Miranda and Casquilho or by Henriques and Coelho.

The implementation procedures are detailed in Miranda (2013, 2014) and the sequential steps are synoptically described in below (the code can be made available for the interested reader upon request).

1. ***New OPL Project***—opening a new workspace for the work at hand, saving dedicated memory and associated files;
2. ***Create Project***—automatically creating files in the workspace, such as, the MILP model, the problem data, and execution configurations; different data and models can be allowed under different execution configurations, for example on stochastic problems;
3. ***Problem model***—in the model file, firstly, sets and parameters are to be sequentially defined; then, in a second step, the positive variables are introduced; finally, after the initialization of all these attributes, the objective function and restrictions of the MILP model are included too;
4. ***Problem data***—in the data file, sets and parameters are specified, then the associated values are introduced; in the case at hand, the activities set, the set of immediate predecessors for each activity, the set of activities with no immediate successors, the upper bound for the project time, the activities duration, and additional data are required;
5. ***Run the solver***—the execution of CPLEX solver follows the default implementation, the usual model and data files are selected;
6. ***Problem solution***—with the proper instruction in the model file, the MILP solution is presented on screen (as in Fig. 10).

By direct observation of Fig. 10, the project completion time is confirmed ($T = 23$), so as the starting times (ST: activities in alphabetic order) for all the critical activities (in bold line, A-B-I-K-L) in the project schedule. This solution corresponds to the overload schedule presented in the first point of Sect. 3 (Fig. 3), allowing this way a better comparison between the computational procedures and the graphic approaches.

Resources restrictions shall apply during all the project period, and a logic-restrictions set on resources availability is usually included in MILP models. The resources restrictions consider the quantity of resource r necessary to perform activity i, R_{ir}, the associated set of binary variables, y, and the sum of the required resources shall not overpass the quantity of resource available at time t, A_{rt}:

Fig. 10 Problem solution for the case project through MILP modeling

```
// solution (optimal) with objective 23
(...)
T = 23;
ST = [0 1 8 1 8 11 19 20 7 18 21 22];
```

$$\sum_{i=1}^{N} R_{ir} \cdot \left(\sum_{k=\max(L_i,\, t-d_i+1)}^{\min(U_i,t)} y_{ik} \right) \leq A_{rt}, \quad t = 1, \ldots, Tmax; \ \forall r \in R \qquad (8)$$

With a sequence of MILPs, not only basic constraints are ensured but also other objectives and types of constraints can be considered. Next to the makespan minimization with no overload on resources (T = 34), a second MILP model is solved to minimize payments within such makespan. Therefore, the results in Fig. 6 and the related developments are also obtained by MILP solving.

5 Conclusions and Final Comments

A simple case of project scheduling is presented, graph-based and heuristics approaches are applied, such as a MILP model is implemented in IBM ILOG CPLEX. The former approaches allow the treatment of small size projects, real world applications on various fields, and they are widely used on practice.

To better provide larger applications, to address uncertainty on activities duration, or even to consider time-cost trade-offs, then commercial software is suggested. Computational implementation of LP or MILP models, friendly software environments, they allow to evaluate different scenarios, to update project data along its time horizon, and also to avoid the burden of step-by-step procedures.

In this sense, to implement a second MILP model to level resources utilization is a challenge that remains open, being the results provided on the final figures. This procedure is well-known, since the computational implementation of mathematical models and the results verification in the open literature allows readers to extend their capabilities beyond either the commercial applications or the bibliographic references.

Acknowledgments Authors thank Universitat Politècnica de Catalunya (UPC) and Portalegre Polytechnics Institute (IPP). The presented works are partially developed with the CERENA/IST support on behalf of FCT project UID/ECI/04028/2013. We also thank Samuel Moniz (INESC/UP, CEG/IST/UL) for the support on IBM ILOG CPLEX lab sessions and the chapter reviewers for their useful comments.

References

Artigues, C., Brucker, P., Knust, S., Koné, O., Lopez, P., & Mongeau, M. (2013). A note on "event-based MILP models for resource-constrained project scheduling problems". *Computers & Operations Research, 40*, 1060–1063.

Burgess, A. R., & Killebrew, J. B. (1962). Variation in activity level on a cyclical arrow diagram. *Journal of Industrial Engineering, 13*, 76–83.

Koné, O., Artigues, C., Lopez, P., & Mongeau, M. (2011). Event-based MILP models for resource-constrained project scheduling problems. *Computers & Operations Research, 38*, 3–13.

Lusa, A. (2012). Workshop on project development. In Erasmus Intensive Programme on "Optimization and DSS for SC", Portalegre Polytechnics Institute, Portalegre, Portugal, July 01–15, 2012.

Lusa, A. (2013). Project development: Integrated aggregate planning. In Erasmus Intensive Programme on "Optimization and DSS for SC", Portalegre Polytechnics Institute, Portalegre, Portugal, July 07–21, 2013.

Miranda, J. L. (2013). Workshop on Project Development—Lab Session. In Erasmus Intensive Programme on "Optimization and DSS for SC", Portalegre Polytechnics Institute, Portalegre, Portugal, July 07–21, 2013.

Miranda, J. L. (2014). Workshop on project development. In Erasmus Intensive Programme on "Optimization and DSS for SC", Portalegre Polytechnics Institute, Portalegre, Portugal, July 06–20, 2014.

IBM ILOG CPLEX Optimization Studio OPL Language User's Manual. (2015a). © Copyright IBM Corporation 1987. http://www-01.ibm.com/support/knowledgecenter/SSSA5P_12.6.2/ilog.odms.studio.help/pdf/opl_languser.pdf?lang=en

IBM ILOG CPLEX Optimization Studio CPLEX User's Manual. (2015b). © Copyright IBM Corporation 1987. http://www-01.ibm.com/support/knowledgecenter/SSSA5P_12.6.2/ilog.odms.studio.help/pdf/usrcplex.pdf?lang=en

IBM ILOG CPLEX Optimization Studio CP Optimizer User's Manual. (2015c). © Copyright IBM Corporation 1987. http://www-01.ibm.com/support/knowledgecenter/SSSA5P_12.6.2/ilog.odms.studio.help/pdf/usrcpoptimizer.pdf?lang=en

Pritsker, A., Watters, L., & Wolfe, P. (1969). Multi-project scheduling with limited resources: A zero-one programming approach. *Management Science, 16*, 93–108.

Sustainability Analysis of Industrial Processes

Henrique A. Matos and Ana Carvalho

Abstract Our planet has been extensively impacted by the enormous consumption of natural resources and due to the high level of emissions coming from the productive systems. This situation is imposing severe burdens to the planet, leading to environmental disturbances and a huge level of pollution, which is causing a significant increase of human diseases. Sustainability Analysis at the industrial process level is mandatory to accomplish the sustainable development among nations. The aim of this chapter is to clearly present the sustainability agenda across the past decades and integrate that concept in the industrial processes analysis. For that, metrics and tools available to assess and improve industrial processes in terms of sustainability are presented. Some examples of the application of these tools are also described.

Keywords Sustainability · Process design · Indicators · Sustainability tools

1 Introduction

Resources are becoming scarce and the planet will not last long with the current levels of consumption, of fossil fuels, minerals, among other resources. The industrial productive system is responsible for the production of products and services, required to meet the demanding standards of a consumerist population. The major problem in this civilization is the use amount of resources extract from the nature, and out of that only 20 % in terms of weight is effectively used (OECD 2012). The remaining resources are wasted in different ways, gaseous emissions, liquid effluents and solid disposals. Due to the aforementioned situation, a strong

H.A. Matos (✉)
CERENA, Instituto Superior Técnico, Universidade de Lisboa, Lisboa, Portugal
e-mail: henrimatos@tecnico.ulisboa.pt

A. Carvalho
CEG-IST, Instituto Superior Técnico, Universidade de Lisboa, Lisboa, Portugal
e-mail: anacarvalho@tecnico.ulisboa.pt

A.P. Barbosa-Póvoa et al. (eds.), *Optimization and Decision Support Systems for Supply Chains*, Lecture Notes in Logistics,
DOI 10.1007/978-3-319-42421-7_12

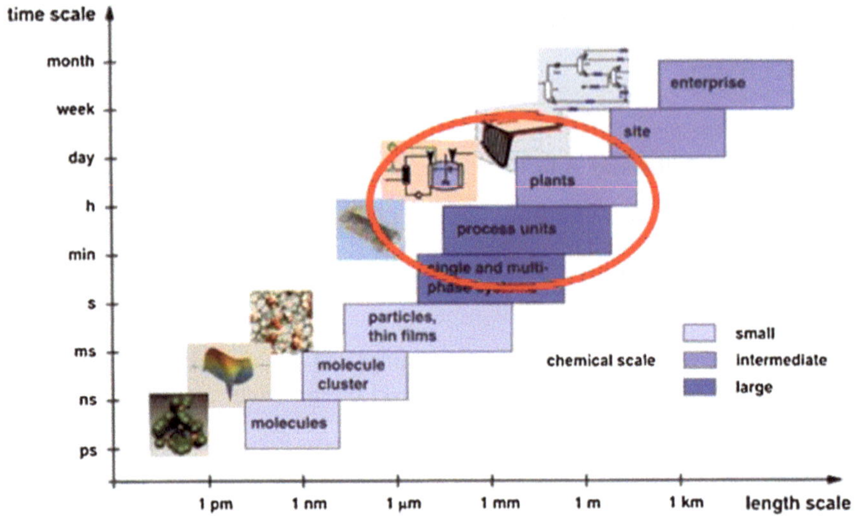

Fig. 1 Multi-scale approach in process systems engineering (WRI 2005)

activity has been carried out in the industrial production sites in order to improve the sustainability of the industrial processes. Among the scientific/technical developments it can be highlighted the new and more efficient catalysts and more efficient separations through the integration of new agents. Moreover, the industrial process intensification and integration attitudes, also move the processes towards the right direction, leading to systems, which seem to have all the plant inside of an environment-friendly container, sometimes called the "Banana Container".

In order to obtain a more efficient and environmental friendly productive industry, a holistic vision, covering a complete multi-scale approach is required. The multi-scale approach spans from the molecule level through all supply chain management, considering sustainable suppliers and final adequate disposal of the non-used items (Fig. 1).

This chapter will be dedicated to a cornerstone of any supply chain, the production process. This chapter aims to present the concept of sustainability and how they should be integrated in the Sustainability Analysis of Industrial Processes. For that purpose the methodologies and indicators suitable for sustainability improvement and assessment at the process level, will be revised. This chapter is organized as follows. In the next section an overview of the sustainability concept will be presented. In Sect. 3 the assessment methodologies for the three pillars of sustainability are described. Then in Sect. 4 the integrated methodologies for process improvement towards sustainability are presented. Conclusions are presented in Sect. 5.

2 Sustainability Analysis Concept

The widespread awakening on the issue of sustainability in the development of society was made during the decade of the 60s of the twentieth century. However, sustainability has only been considered a landmark, achieving the public recognition, with the publication of the book Limits to Grow in 1972 by Meadows et al. (1972). This book was the result of a group of industrialists who gathered in Rome to discuss the "new" international problems. This work describes the results of a computer model of human evolution ("World 3"), which identifies the consequences for the planet, of the current exponential industrial development and population growth. The book's conclusions are something catastrophic, and are based on scenarios away from the reality of a finite world. These conclusions had a strong impact on public opinion and especially in political power.

Later, in 1992, the same research group published an updated work through the book Beyond the Limits (Meadows et al. 1992), and the findings were similar to the previous edition, which showed that much work was still required. The concept of Sustainability was formally introduced by the World Commission on Environment and Development (WCED) of the UN headed by the Prime Minister of Norway, Gro Harlem Brundtland in the report "Our Common Future" or Brundtland Report (Brundtland 1987). In this report sustainability has been defined as: "meet the needs of the present generation without affecting the ability of future generations get their supply." The report indicates a number of measures that should be taken by countries to promote sustainable development. One of the proposed measures, in the Brundtland report, concerns the need for the United Nations (UN) to implement a sustainable development program. This recommendation of the commission led to the development of Agenda 21 that began in 1989 with the approval of a special meeting with the United Nations, in a conference on environment and development. All member states of the UN developed a complex process of review, consultation and negotiation, culminating in the Second United Nations Conference on Environment and Development, known as Rio-92 or Eco 92, held in Brazil. There, representatives of 179 governments agreed to adopt a deep programme on the subject Agenda 21 had a close monitoring from which were made adjustments and revisions. The first step was the Rio+5 conference in 1997, at UN Headquarters in New York; later with the adoption of an additional calendar called the Millennium Development Goals (Millennium Development Goals), with particular emphasis on globalization policies and the eradication of poverty and hunger, adopted by 199 countries at the 55th UN General Assembly, held in 2000 at New York; The Declaration of the Millennium Development Goals outlined eight basic objectives, which are (Almeida, 2007): (1) Eradicate extreme poverty and hunger; (2) Achieve universal primary education; (3) Promote gender equality and empower women; (4) Reduce child mortality; (5) Improve maternal health; (6) Combat HIV, malaria and other diseases; (7) Ensure environmental sustainability; (8) Develop a global partnership for development.

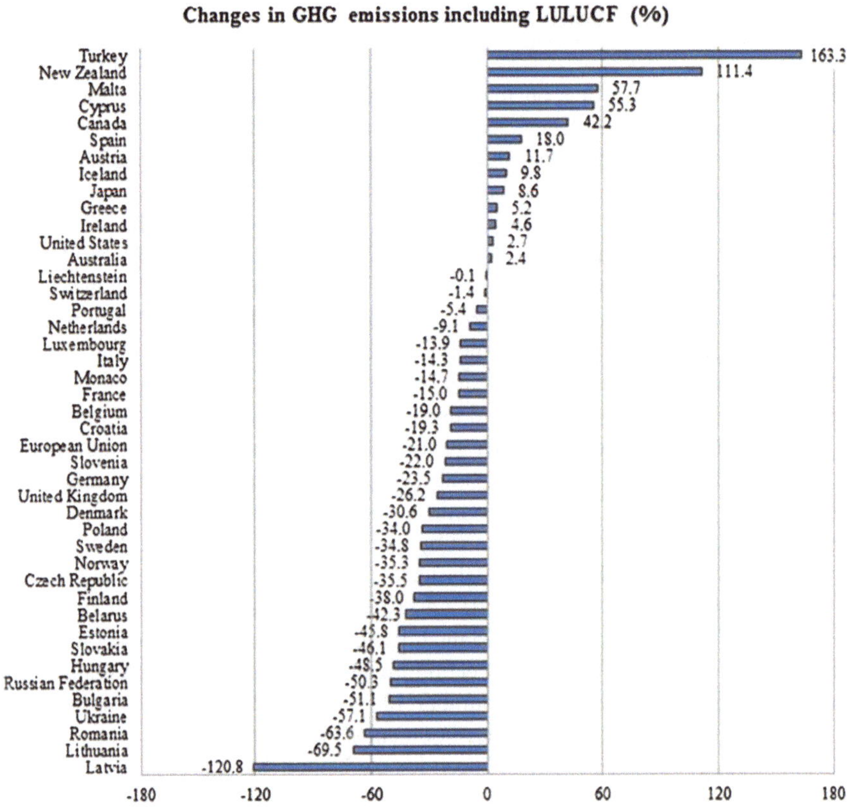

Fig. 2 Total aggregate greenhouse gas (GHG) emissions of individual parties including Land Use, Land-Use Change and Forestry (LULUCF), 1990–2012 (UNFCCC 2015)

Meanwhile, the Kyoto Protocol is an international agreement linked to the United Nations Framework Convention on Climate Change, which commits its Parties by setting international binding emission reduction targets. Recognizing that developed countries are principally responsible for the current high levels of GHG emissions in the atmosphere as a result of more than 150 years of industrial activity. During the first commitment period, 37 industrialized countries and the European Community committed to reduce GHG emissions to an average of 5 % against 1990 levels. During the second commitment period, Parties committed to reduce GHG emissions by at least 18 % below 1990 levels in the eight-year period from 2013 to 2020; however, the composition of Parties in the second commitment period is different from the first as shown in Fig. 2 where the GHG are accounting including Land Use, Land-Use Change and Forestry (LULUCF).

Some other Summits on Climate Changes and Sustainable Development were carried out such as the Summit in Johannesburg, South African in September 2002,

Copenhagen in 2009, Cancun in 2010 and Rio+20 in 2012. All occur to establish more targets and assess the countries performance.

In 25–27 of September, 2015 the United Nations Development Summit defined the new Agenda 2030, which is a plan of action for people, planet and prosperity. It also seeks to strengthen universal peace in larger freedom. Eradicating poverty in all its forms and dimensions, including extreme poverty, is the greatest global challenge and an indispensable requirement for sustainable development. All countries and all stakeholders, acting in collaborative partnership, will implement this plan. The new 17 Sustainable Development Goals (SDG's) and 169 targets, which are announced at that summit demonstrated the scale and ambition of this new universal Agenda. The SDG's and targets will stimulate action over the next fifteen years in areas of critical importance for humanity and the planet:

People	To eradicate the poverty and hunger, in all their forms and dimensions, and to ensure that all human beings can fulfil their potential in dignity and equality and in a healthy environment.
Planet	To protect the planet from degradation, including sustainable consumption and production, sustainable managing its natural resources and taking urgent actions on climate change, so that it can support the needs of the present and future generations.
Prosperity	To ensure that all human beings can enjoy prosperous and fulfilling lives and that economic, social and technological progress occurs in harmony with nature.
Peace	To foster peaceful, just and inclusive societies, which are free from fear and violence. There can be no sustainable development without peace and no peace without sustainable development.
Partnership	To mobilize the means required to implement this Agenda through a revitalised Global Partnership for Sustainable Development, based on a spirit of strengthened global solidarity, focussed in particular on the needs of the poorest and most vulnerable and with the participation of all countries, all stakeholders and all people.

The concept of sustainable development must be assimilated by the leaders of the companies as a new way to produce without degrading the environment, extending this culture at all levels of the organization. A systematic analysis of the company's production process in terms of impact on the environment, economic and social aspects should be conducted. This analysis should result in the implementation of a project, which combines production, social and environmental preservation, with technology adapted to that principle (Ramôa Ribeiro 2009). Sustainability Analysis of Industrial Processes emerges as an indispensable practice to design and evaluate new processes and/or existing ones (retrofit design). It is essential to assess the existing process and propose new design alternatives in terms of the three pillars of sustainability. The three main components of sustainability analysis in industrial processes are accordingly to its basic definition the following:

- Stability and cost-effectiveness: should be sought technological solutions that lead to optimum operating point at the level of minimum total cost or maximum profitability;
- Ecological balance: by including the rational use of raw materials, the preservation of natural ecosystems and mitigating the effects of climate change;
- Social development and equity: adapting the existing rule to the concrete situation, observing the criteria of safety, justice and equality.

3 The Sustainability Pillars and Their Assessment

Sustainability is comparable to a tripod where each leg corresponds to a specific pillar: Economic, Environmental and Social. It makes no sense to walk in each of the fields independently, but instead these fields should be harmonized in order to achieve an appropriate outcome. To attain this goal, it is required to have methodologies and tools, which deal with the three pillars of sustainability. It is also required to incorporate sustainability indicators, covering the different areas, so that industrial processes can be assessed.

3.1 Methodologies and Tools to Address the Economic Pillar

Several approaches could be used to successful employ the sustainability economic pillar, however in the industrial process studies the best practice is through process modelling and optimization. The descriptive mathematical models of the processes, through equality or inequality constraints is added to an economic objective function usually based on operating costs or profit. A possible technical-economic solution should be found, corresponding to the best value for a given objective. This means the optimal operational conditions are set to the variables, leading to the minimum cost or maximum profit.

Several commercial software packages with a modular structure or individual components are used to solve the representative model resolution sequentially (e.g. ASPEN Plus, PRO-II or HYSYS). Some illustrative of the Aspen Suite interface and results are shown in Fig. 3.

Alternatively, a simultaneous resolution of the mathematical model could be chosen. Some models equation-based written in language as *gPROMS* (http://www.psenterprise.com/company.html) can be used to describe process steady–state or dynamic behaviour. The interface could be illustrate in Fig. 4.

GAMS—General Algebraic Modeling System (http://www.gams.com/) is another language that could provide as an interface to connect the model with the different solvers. The model of the process could be well described by a linear or

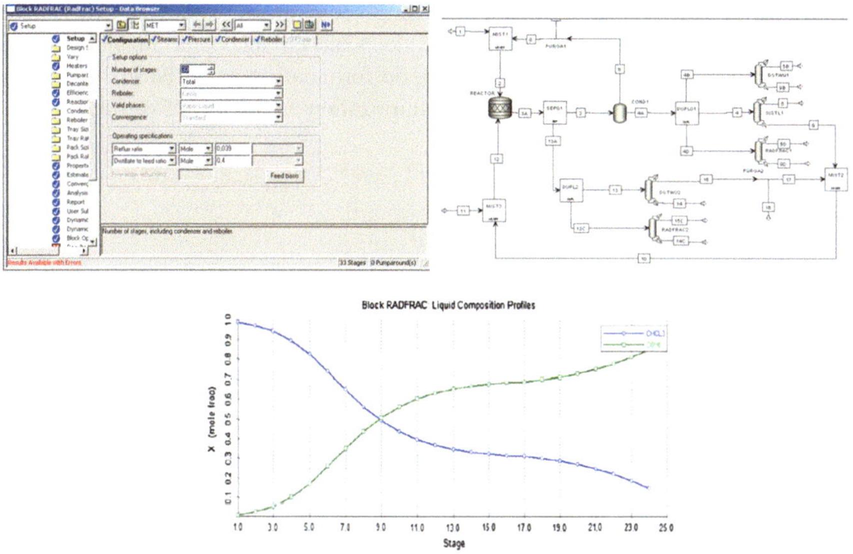

Fig. 3 Aspen Suite interface and results as Process Simulator

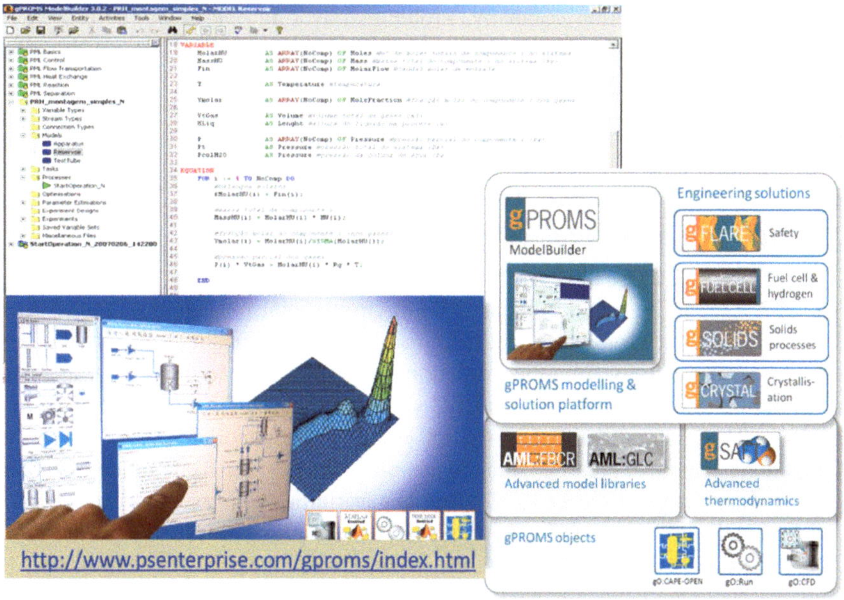

Fig. 4 gProms interface and some illustrative outputs

non-linear mathematical formulation using continuous variables (x) and/or discrete/binary ones (y) sometimes called decision variables and transforming the problems in mixed formulation. The generic formulation could be given by considering the optimization direction the minimization as the objective function (f):

Minimise $\boldsymbol{f}(\boldsymbol{x}, \boldsymbol{y})$

Subject to $\boldsymbol{h}(\boldsymbol{x},\boldsymbol{y}) = 0$

$\boldsymbol{g}(\boldsymbol{x},\boldsymbol{y}) \leq 0$

x Continuous variables (x_1, x_2, x_3 ... x_n)

y Binary variables (y_1, y_2, y_3 ... y_m)

As an example of this last approach, are the challenging problems in real world scenarios on the planning and scheduling of the complex oil supply chain. There is a growing need to establish decision support tools capable of dealing with competition within this industry and enable the replacement of simple procedures/tools, such as spreadsheets, with more efficient ones. Relvas et al. (2007) studied a medium term scheduling problem of a pipeline and the associated end-of-pipe tank farm and modelled it as a mixed integer linear problem (MILP). This model represents the articulation between the pipeline schedule and the inventory management at the tank farm. The improvements include variable flowrate, pipeline stoppages, and variable settling period with product. On this scheduling model, a novel rescheduling methodology is developed, taking into account the variability of real plant changes with the definition of revisions of schedules in an effective way. The model and respective rescheduling strategy have been applied to a scenario of a Portuguese oil distribution company. Problems covering one month of time horizon, including initial plans and their revisions, with more than one perturbation, have been successfully solved.

3.2 Methodologies and Tools to Address the Environmental Pillar

The LCA—Life Cycle Assessment is an environmental management tool, which allows the identification and quantification of the environmental impact of a product, process or activity from the extraction of raw materials to the placing as waste in a landfill (cradle to grave). The LCA has been used in numerous industrial sectors such as energy, and today is a well-established methodology with numerous applications in industry, research and decision policies. This methodology is based on four main steps (ISO 14040 2006): (1) Definition of objectives and scope; (2) Life Cycle Inventory; (3) Life Cycle Impact Assessment; (4) Interpretation.

Several methods to evaluate the environmental process impact assessment of industrial chemical processes have been presented in the literature. Although these methods follow the same steps (usually, characterisation, normalisation and weighting), each method addresses different impact categories, considering specific nomenclatures and using different taxonomies to classify their outputs. When

LCIA- Life Cycle Impact Assessment

- Methods oriented to the Problem (Mid-Point Indicators)
  - CML – **Heijungs, R,** J.B. Guinée, G. Huppes e R.M. Lankreijer. *Environmental Life Cycle Assessment of Products: Guide.* Report 9266, **Leiden, The Nederland's : Centre of Environmental Science** (CML), **1992**
  - EDIP – **Hauschild, M.** e H. Wenzel. *Environmental Assessment of Products, vol.2:Scientific Background.* London: Chapman and Hall, **1998**.
- Methods oriented for Environmental Damage (End-Point Indicators)
  - EPS - Steen, B. "A Systematic Approach to Enviromental Prority Strategies in Product Development (EPS)." CPM report , Chalmers, **1999**.
  - Eco-Indicator 99 - **Goedkoop, M.**, e R. Spriensma. *The Eco-Indicator 99: A damage Oriented Method for Life Cycle Assessment.* Methodology Report, Amersfoort, The Netherlands: **Pré-Consultants, 2001**.
- Mixed Methods :
  - IMPACT 2002^{+} - **Jolliet, O.**, et al. "IMPACT 2002+: a new life cycle impact assessment methodology." *International Journal of LCA* 8(6) (**2003**): 324-330.
  - **ReCIPe 2008** - **Goedkoop, M., Heijungs, R.,** Huijbregts,M., An De Schryver, Struijs, J., van Zelm, R. *ReCiPe 2008 -A life cycle impact assessment method which comprises harmonized category indicators at the midpoint and the endpoint level, **2009***

Fig. 5 Methods to carried out LCIA—Life Cycle Impact Assessment

industrial processes are being assessed by several methods, different results are obtained. The comparison of these non-standardised results turns out to be a difficult task. The careful selection of the indicators used in this step should be the subject of a thorough analysis and study (Dewulf and van Langenhove 2006; Carvalho et al. 2014). In Fig. 5 the most relevant methods to carry out LCIA are summarized.

A preliminary study of the different environmental process impact assessment methods revealed important differences in the type of output, namely on the different impact categories. For example, CML method presents 14 midpoint indicators (Heijungs et al. 1992; Guinée et al. 2001), whereas Eco-indicator presents 3 endpoint indicators (Goedkoop and Spriensma 2001), which are combined into a final indicator. For instance, according to the proposed terminology, the ReCiPe method's (Goedkoop et al. 2009) output includes 18 indicators that quantify 18 impact categories, which are designated by the authors as midpoint impact categories/indicators. Another mixed method is the IMPACT 2002+ was introduced by Jolliet et al. in 2003.

The Environmental Analysis processes can also be carried out based on a prospective calculation method of environmental impact (PEI—Potential Environmental Impact), which was introduced by EPA—Environmental Protection Agency (1995). The PEI is a relative measure of the effect that a chemical has potentially on human health and the environment. A mathematical algorithm called WAR-Waste Reduction was developed to determine the value of global PEI process. The result of PEI is a process for calculating a measure of the impact of the

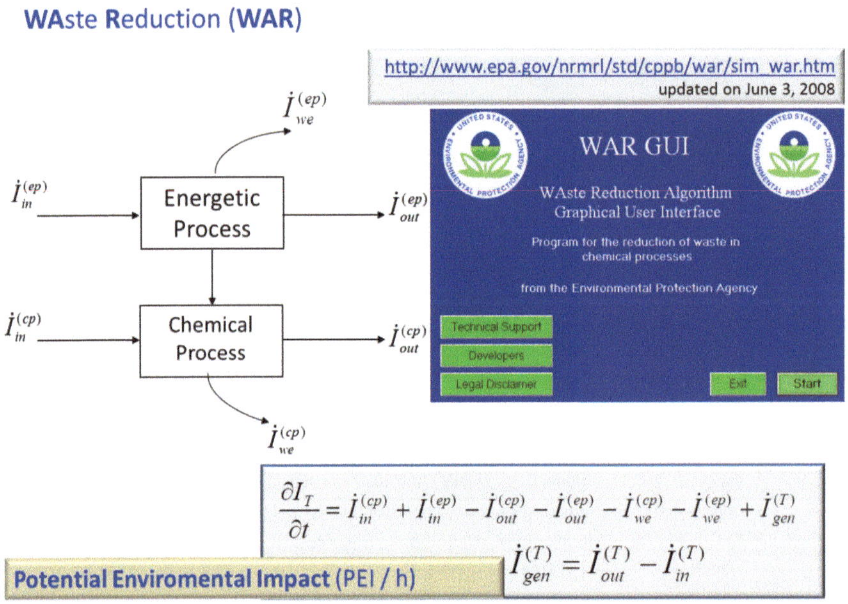

Fig. 6 WAR—Waste Reduction framework (adapted from Cabezas et al. 1997)

process effluent (Cabezas et al. 1997). The purpose of the application of the methodology is to get the characterization of the process, so that you can make the comparison between possible alternatives (structural and parametric) for minimizing the environmental impact and that will have a direct influence on the choice of alternative leading to a global minimization of effluent. These indices include eight different categories of impacts and can be calculated using the Process Simulation Excel (a first step) and Aspen Plus (a second step).

The eight categories used to calculate the environmental impact potential (PEI) through the WAR algorithm are divide in two groups (Fig. 6): Atmospheric and Toxicological Impacts (Cabezas et al. 1997).

In the group of **Atmospheric Impacts** four categories could be pointed out:

GWP (Global Warming Potential): measures the absorption capacity of infra-red radiation of a mass unit of a given substance, compared to a mass unit of CO2, over an established time of 100 years.

ODP (Ozone Depletion Potential): compares the reaction rate between a mass unit of a given substance with ozone, and a mass unit of CFC-11 (trichloro-fluoromethane) with ozone, to originate molecular oxygen.

PCOP (Photochemical oxidation Potential): compares the reaction rate between a mass unit of a given substance with the hydroxyl radical (OH.), and a mass unit of ethylene with the same radical.

AP (Acidification Potential): quantifies the release rate of H+ in the atmosphere of a given substance, compared to the release rate of H+ due to the formation of SO2.

In the group of **Toxicological Impacts** four categories could be pointed out:

HTPI (Human Toxicity Potential by Ingestion): based on LD50 for mice (mg of substance/mg of mouse), because most of these values for humans is not available. When LD50 for mice is unknown, estimations are made using molecular methods.
HTPE (Human Toxicity by exposure—dermal and inhalation): calculated based on TWA (Time-Weighted Averages) and TLV (Threshold Limit Values) published annually by OSHA, ACGIH and NIOSH.
TTP (Terrestrial Toxicity Potential): It is obtained using the same parameters used to calculate HTPE.
ATP (Aquatic Toxicity Potential): Based on LC50 (similar to LD50) of the aquatic specie Pimephales promelas.

3.3 Methodologies to Address the Social Pillar

The social pillar still comes up with low development in terms of the engineering community, however it can be approached in a pragmatic way, as a strand of health effects and intrinsic safety of processes. For Industrial Processes a special attention is given to Health and Inclusive Design. The human health risk associated with the chemical product is dependent on the speed at which the product is released and the severity of product effects on the environment. The exposure estimation methods can be divided into two classes: Occupational and Community. Occupational exposure occurs at the workplace due to possible use or production of harmful chemicals. The exposure to chemicals can occur by inhalation in the working space, by ingestion of contaminated food or powder, or by direct contact of the substance with the skin or eyes. In Community terms the most common form of exposure is by inhalation of the surrounding air or by direct or indirect use of contaminated water (Allen and Shonnard 2002).

Several Indices could be pointed out to characterise the industrial safety based on different methodologies and indicators:

- Dow Fire and Explosion Index (Dow/AIChE 1987)
- Dow Chemical Exposure Index (Dow/AIChE 1998)
- Mond Index (Tyler 1985)
- Hazop—Hazard and Operability Analysis (CCPS 2009)
- PIIS—Prototype Index of Inherent Safety (Edwards and Lawrence 1993)
- ISI—Inherent Safety Index (Heikkila 1999)
- I2SI—Integrated Inherent Safety Index (Khan and Amyotte 2004)
- PRHI—Process Route Healthiness Index (Hassim and Edwards 2006)

The Inherent Occupational Exposure index (PRHI—Process Route Healthiness Index) was developed to quantify the risk associated to the exposure of workers to industrial processes. This index is influenced by the impact on health due to the potential releases of chemicals (small leaks and fugitive emissions) and the concentration in the air inhaled by workers. The index is calculated based on five indicators:

- ICPHI—Inherent Chemical and Process Hazard Index;
- HHI—Health Hazard Index;
- MHI—Material Harm Index;
- WEC—Worker Exposure Concentration;
- OEL—Occupational Exposure Limit.

Hassim and Edwards (2006) proposed a calculation routine to deliver the various indicators that lead to the calculation of PRHI index.

$$PRHI = ICPHI \times MHI \times HHI \times \frac{WEC_{\max}}{OEL_{\min}}$$

This routine can be applied to cases which present different reaction pathways helping in decision making.

The Inclusive Design is a key approach to obtain an Inherent Safety design (CCPS 2009). It is related to four strategies:

- **Minimize**: the use of less hazardous substances;
- **Replacement for Change**: replacing the material by other less dangerous;
- **Moderate**: carrying out the processes under less dangerous conditions and minimizing the impact in the site if an eventual release of hazardous material or energy occurs;
- **Simplify**: creating simple design facilities in order to reduce complexity and operations mistakes.

For instance to follow the aforementioned **Minimize** strategy, Process Intensification (PI) appears as a possible solution. PI consists of the development of novel apparatuses and techniques that, compared to those commonly used today, are expected to bring dramatic improvements in manufacturing and processing, substantially decreasing equipment-size/production-capacity ratio, energy consumption, or waste production, and ultimately resulting in cheaper, sustainable technologies. Figure 7 shows an example of the minimize strategy by PI of the methylacetate process.

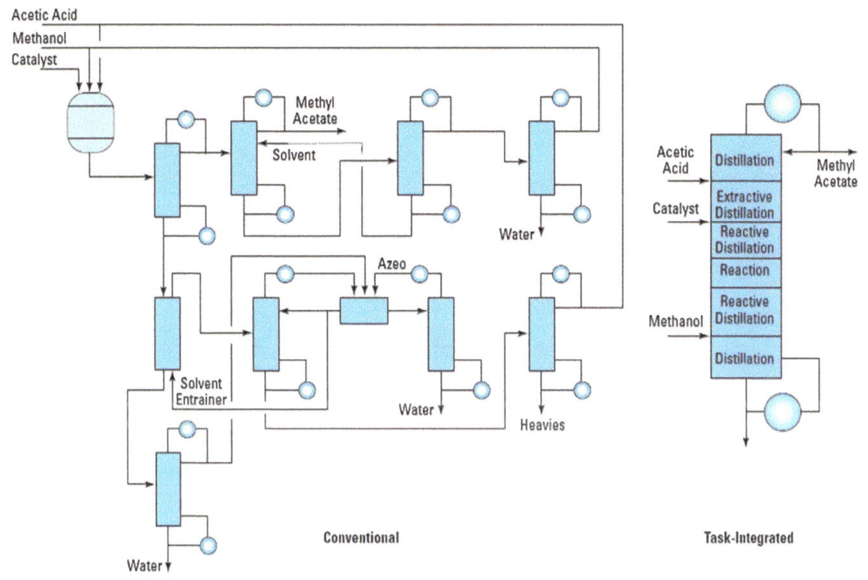

Fig. 7 Example of Inherent Safety design using Process Intensification (drawing courtesy of Eastman Chemical, CCPS 2009)

3.4 Incorporating and Assessing Sustainability in Industrial Processes

Several frameworks have been presented in order to assess sustainability, some at a national level and some at a process level. For instance, General Sustainability Indicators set for the United Kingdom (Defra 2008) consisted of 68 indicators comprising 126 measures covering four major areas: (1) Sustainable consumption and production; (2) Climate change and Energy; (3) Protecting natural resources and enhancing the environment; (4) Creating sustainable communities and a fairer world. In June 2013, based on one consultation it was created a new Sustainable Development Indicators set (SDIs). The new set provides (Defra 2008) an overview of national progress towards a more sustainable economy, society and environment. This SDIs are used as a means of assessing whether the nation as a whole is developing sustainability, and as a means for policy-makers to identify more sustainable policy options. As it can be seen, one of the most critical area in the supply chain is the production, since this entity will influence the four main areas of sustainability at the national level. Therefore, these national frameworks have been adapted to Industrial Analysis. One of the most known specific frameworks are the Sustainable Development Progress Metrics recommended by the Institution of Chemical Engineers for use in Process Industry (IChemE 2007). The three metrics related to the sustainability pillars have some indicators such as: Resource usage (Energy, Raw materials, water, land); Emissions, effluents and waste (atmospheric

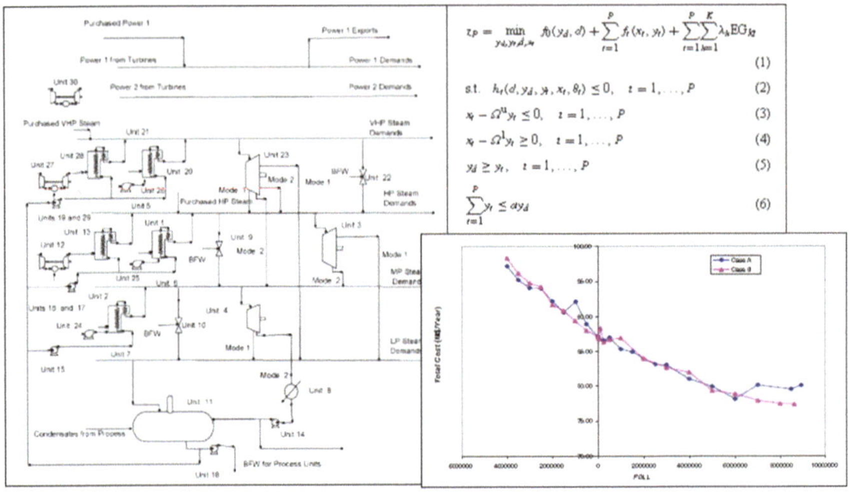

Fig. 8 Utility Network, mathematical model and Pareto Front

impacts, Aquatic impacts and impacts to land); Profit, value and tax (Investments, Additional economic items); Workplace, Society and additional social items.

The previous metrics can be incorporated in the economic models presented in Sect. 3.1, leading to a complete integration of sustainability into process analysis. The balance between the environmental impacts and other economic and social issues leads to multi-objective problems (Young et al. 2000). Pareto front strategy is usually applied to find an integrated process with a clear reduction of environmental impact and to maintain the economic capacity to the required profitability, considering the social effects are not changed. The Pareto front corresponds to a set of solutions where an improvement in one of the objectives can only be achieved by increasing the value of another object. It is a decision support tool for situations where the weights for each objective function or relation between them can be a difficult task (Young and Cabezas 1999; Sikdar and El-Halwagi 2001). An illustrative example is given in Fig. 8 where a utility network is designed assuming the minimization of Total Cost and the reduction of the emissions (Francisco and Matos 2004).

4 Integrated Methodologies

During the past decades several methodologies aiming to analyse sustainability in industrial processes have been presented in the literature. These methodologies have as the ultimate goal the reduction of material consumption, energy, water, environmental and social impacts. These methodologies can be summarized into two types:

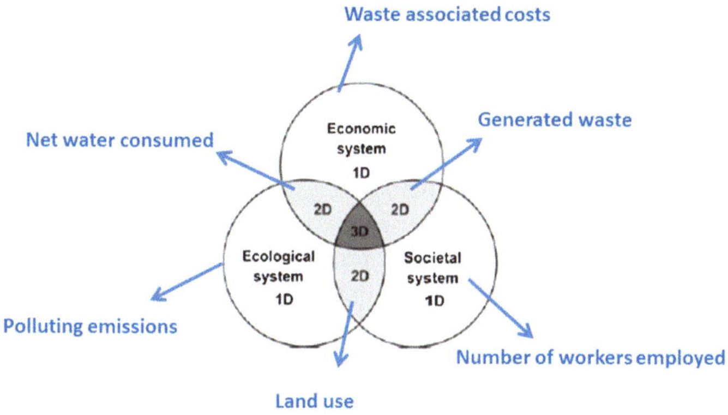

Fig. 9 Example of different dimensions of 3D sustainability analysis (Martins et al. 2007)

(i) Methodologies integrating sustainability through a set of indicators and aiming to assess process alternatives, helping the decision maker in the choice of the best alternative;
(ii) Methodologies which evaluate the process in terms of Sustainability, pointing out the critical points and generating new design alternatives. An final assessment is also performed to compare the design alternatives;

The first type of methodologies can be illustrated by the work developed by Martins et al. (2007). This work presents a conceptual structure based on indicators classified in levels, covering three dimensions (3D), two dimensions (2D) and one dimension (1D) indicator. The analysis is performed initially by calculating the 3D windows for different alternatives obtained heuristically and when their analysis have not been conclusive it requires the calculations of other 2D or 1D indicator's (Fig. 9).

The sustainability metric proposed by Martins et al. (2007) for 3D Analysis is based on four indicators:

- **Energy Intensity**: Ratio between the total non-renewable energy consumption and the quantity of produced product;
- **Materials Intensity**: Ratio between the quantity of raw materials, solvents and other substances and the quantity of produced product;
- **Potential Chemical Risk**: Related to the process dangerousness to the human health due to the use, storage and manipulation of the products;
- **Potential Environmental Impact**: Refers to the environmental impact related to the emissions and discharge of hazardous chemicals to the environment.

The framework proposed to obtain the four indicators has been applied to a simple example of solvent selection, analysing the possibility of replacing the extracting solvent benzene by methyl-n-pentyl ether. The calculations based on process simulation to obtain the energy and materials intensity and safety class tables to obtain the Potential Chemical Risk and Environmental Impact could provide a final result of improvement in all the indicators if the process/solvent alternative is used.

The second type of integrated methodologies is illustrated through the methodology developed by Carvalho et al. (2008).

This methodology has been developed for analysing sustainability in industrial processes. The methodology identifies the critical points of the process to change (operating equipment, change materials, etc.). The methodology could guarantee in the end a more sustainable process alternative according to economic, environmental, social and safety issues. The method uses mass and energy indicators, which screen the process, allowing the identification of process bottlenecks, this means the process point that presents mass losses, high costs or undesirable accumulation of species. The indicators that have more severe problems (i.e. high potential for improvement) are selected. A sensitivity analysis to identify target windows, as well as target variables to the desired improvement is performed. The method further involves the use of process synthesis algorithms to suggest new process alternative(s).

The new alternatives are assessed in terms of indicators and metrics. A comparison is carried out between one new alternative and the initial process (base case). In this evaluation metrics for sustainability and security indexes (endpoints) set by IChemE (2007) are used. If the indicators remain constant or improve parameters or, the new alternative is accepted as a more sustainable alternative; otherwise a new process alternative should be assessed and evaluated. This methodology has been incorporated in a software tool, called *SustainPro* (Carvalho et al. 2013). The tool Excel-based has made this methodology a systematic and fast approach to create new sustainable design alternatives. Figure 10 shows the overview of the SustainPro methodology and the usage of different simulations or characterization tools.

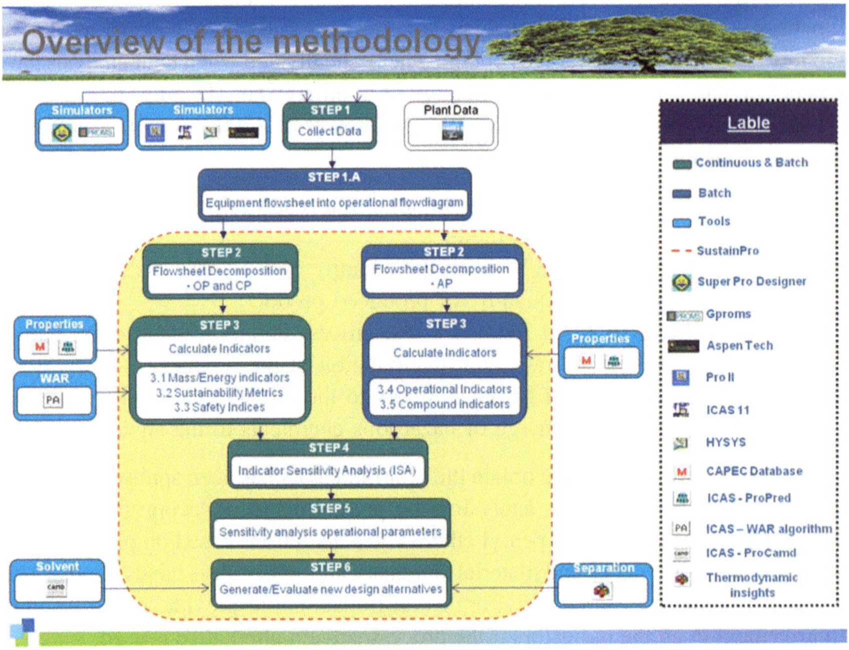

Fig. 10 SustainPro methodology overview with the tools interaction

SustainPro imports input data and performs all the steps of the methodology, as well as the process sustainability assessment for the base case and afterwards for the alternatives. The atmospheric, terrestrial and aquatic impact of the process is determined using the WAR algorithm proposed by the EPA. Social metrics take into account the situation of employees in the workplace, health and safety at the workplace and in society in general. The methodology has been applied to some industrial processes (Carvalho 2009).

One example has been the VCM (vinyl chloride monomer) production The VCM process (PROII 1992) consists in the following sections: (1) Direct chlorination of ethylene. Part of the raw material (1,2-dichloroethane—EDC) is produced by the reaction of ethylene with chlorine; (2) Part of EDC is produced by oxychlorination of ethylene with the recovered acid chloride (HCL) and oxygen; (3) EDC coming from the previous sections and recycled from the process presents approximately 20 compounds as impurities, which should be removed, in a purification process of EDC; (4) Thermal cracking of EDC to form VCM and HCL; (5) VCM's purification.

SustainPro has been applied to the aforementioned process and the most critical areas have been identified. Based on the indicators analysis and on the subsequent sensitive analysis, the recycling of the raw material, turned to be the most critical bottleneck, due to the excess of impurities. SustainPro has applied the synthesis methods, proposing the investment on a membrane, which would be permeable to the impurities, recovering the EDC with a higher purity. The inclusion of this new operation, led to an improvement in the economic pillar (profit increased 0.25 %), in the environmental pillar with the improvement of water and energy metrics (2 %) and finally the social aspects were kept constant, since the safety index has maintained its value from the base case to the new design alternative.

5 Conclusions and Future Work

Sustainability Analysis of Industrial Process is one response of Process Systems Engineering community to the general societal challenge to improve the natural resources usage efficiency. Different approaches of individual characterization of industrial systems (indicators, metrics and ratios) of each of its economic, environmental and social aspects have been properly developed by the scientific community. In a subsequent step, the scientific and technological community elects business areas of engineering knowledge for effective results on the ground of greater equity and sustainability. New methodologies that integrate Sustainability Analysis have been presented, emphasising on new methodologies for identification algorithms or search for critical parts of the process to identify alternatives. These should be characterized in terms of the metrics available so that they can later be recognized as more sustainable alternatives. The presented methodologies and examples point out a positive outlook and constructive approach aimed to generate discussion and develop decision support mechanisms in the conversion of industrial units. There are also job opportunities in the improvement of both the individual techniques for increasing the energy efficiency of processes, either in the development and improvement of new technologies that enable the reduction of energy and/or water consumption. Moreover, two issues still with great potential for research and action for development the integrated methodologies, particularly in the area of operations management and social pillar. The incorporation of knowledge and concepts associated with the vast area of the management of operations, which includes the various methods of obtaining planning and scheduling optimal production should lead to an analysis of wider sustainability. The main goals of this area should be translated into indicators that characterize industrial processes. Their economic and social dimension (e.g. customer satisfaction, increase work efficiency of workers, etc.) should be incorporated into the underlying sustainability metrics for analysis and decision to choose between alternatives. The most intense and deep exploration of the effects on society of certain procedural change or for building integration (e.g. district heating and cooling) can also be further and in this case provides for greater interaction with researchers from the area of sociology. Once developed this aspect, it is supposed that new indicators may be incorporated into the sustainability metrics for complete methodology. Finally, it can be suggested by extrapolation that this approach has some potential to be considered at a macroscopic level of a particular region or country, after the necessary adaptations in the sustainability studies or analysis.

References

Allen, D. T., & Shonnard, D. R. (2002). *Green engineering—Environmentally conscious design of chemical processes*. Englewood Cliffs: Prentice Hall.

Almeida, F. (2007). *Os Desafios da Sustentabilidade—uma ruptura urgente*. Amsterdam: Campus-Elsevier.

Brundtland, G. H. (1987). ONU: Our Common Future Report. Acedido em Julho de 2009. http://www.un-documents.net/wced-ocf.htm

Cabezas, H., Bare, J. C., & Mallick, S. K. (1997). Pollution prevention with chemical process simulators: The generalized waste reduction (WAR) algorithm. *Computers & Chemical Engineering, 21*, S305–S310.

Carvalho, A. (2009). Design of sustainable chemical processes: Systematic retrofit analysis generation and evaluation of alternatives. Ph.D. thesis, IST.

Carvalho, A., Gani, R., & Matos, H. A. (2008). Design of sustainable chemical process: Systematic retrofit analysis and evaluation of alternatives. *Process Safety and Environment Protection, 86*, 328–346.

Carvalho, A., Matos, H. A., & Gani, R. (2013). SustainPro—A tool for systematic process analysis, generation and evaluation of sustainable design alternatives. *Computers and Chemical Engineer, 50*, 8–27.

Carvalho, A., Mendes, A. N., Mimoso, A. F., & Matos, H. A. (2014). From a literature review to a framework for environmental impact assessment index. *Journal of Cleaner Production, 64*, 36–62.

CCPS. (2009). Inherently safer chemical processes: A life cycle approach (2nd ed., p. 411). Wiley-AIChE.

Defra. (2008). *The sustainable development indicators in your pocket.* Defra Publications.
Dewulf, J., & van Langenhove, H. (2006). *Renewables-based technology: Sustainability assessment.* New York: Wiley.
Dow Chemicals. (1998). Dow's chemical exposure index guide. New York, USA: AIChE.
Dow Fire & Explosion Index. (1987). Hazard classification guide (6th ed.). Midland, MI: Dow Chemical.
Edwards, D. W., & Lawrence, D. (1993). Assessing the inherent safety of chemical process routes: Is there a relation between plant costs and inherent safety? *Trans IChemE, Part B, Process Safety and Environmental Protection, 71*(B), 252–258.
EPA. (1995). *Compilation of air pollutant emission factors, Volume I: Stationary point and area sources.* Office of Air Quality Planning and Standards, U.S. Environmental Protection Agency.
Francisco, A. O., & Matos, H. A. (2004). Multiperiod synthesis and operational planning of utility systems with environmental concerns. *Computers & Chemical Engineering, 28*(5), 745–754.
Goedkoop, M., Heijungs, R., Huijbregts, M., De Schryver, A., Struijs, J., & van Zelm, R. (2009). ReCiPe 2008—A life cycle impact assessment method which comprises harmonised category indicators at the midpoint and the endpoint level.
Goedkoop, M., & Spriensma, R. (2001). *The Eco-indicator 99: A damage oriented method for life cycle impact assessment: Methodology report.* Amersfoort: PRé Consultant.
Guinée, J. B., Gorrée, M., Heijungs, R., Huppes, G., Kleijn, R., Koning, A., et al. (2001). Life cycle assessment: An operational guide to the ISO standards. Centre of Environmental Science, Leiden University, CML.
Hassim, M. H., & Edwards, D. W. (2006). Development of a methodology for assessing inherent occupational health hazards. *Process Safety and Environmental Protection, 84*(B5), 378–390.
Heijungs, R., Guinée, J. B., Huppes, G., & Lankreijer, R. M. (1992). Environmental Life Cycle Assessment of Products: Guide. Report 9266, Leiden, The Netherlands: Centre of Environmental Science (CML).
Heikkila, A.-M. (1999). Inherent safety in process plant design—An index-based approach. Ph.D. thesis, VTT Automation, Espoo, Finland.
IChemE. (2007). *A roadmap for 21st century chemical engineering.* Londres: IChemE.
ISO 14040. (2006). Environmental management—Life cycle assessment—Principles and framework. International Organization of Standardization.
Jolliet, O., Margni, M., Charles, R., Humbert, S., Payet, J., Rebitzer, G., et al. (2003). IMPACT 2002+: A new life cycle impact assessment methodology. *International Journal of LCA, 8*(6), 324–330.
Khan, F. I., & Amyotte, P. R. (2004). Integrated Inherent Safety Index (I2SI): A tool for inherent safety evaluation. *Process Safety Process, 23*, 2.
Martins, A. A., Mata, T. M., Costa, C. A. V., & Sikdar. S. K. (2007). Framework for sustainability metrics. *Industrial & Engineering Chemistry Research, 46*, 2962–2973.
Meadows, D. H., Meadows, D. L., & Randers, J. (1992). *Beyond the limits.* USA: Chelsea Green Publishing Co.
Meadows, D. H., Meadows, D. L., Randers, J., & Behrens, III, W. W. (1972). *The limits to growth.* London: Earth Islands.
OECD. (2012). Material resources, productivity and the environment. www.oecd.org/greengrowth. Accessed in September 2015.
PRO/II. (1992). Casebook, Simulation Sciences Inc.
Ramôa Ribeiro, F. (2009). *A Energia da Razão- Por uma Sociedade com menos CO2.* Gradiva.
Relvas, S., Matos, H. A., Barbosa-Póvoa, A. P., & Fialho, J. (2007). Reactive scheduling framework for a multiproduct pipeline with inventory management. *Industrial Engineering and Chemistry Research, 46*(17), 5659–5672.
Sikdar, S. K., & El-Halwagi, M. (2001). *Process design tools for the environment.* London: Taylor & Francis.
Tyler, B. J. (1985). Using the mond index to measure inherent hazards. *Plant/Operations Progress, 4*(3), 172–175.
UNFCCC. (2015). http://unfccc.int/2860.php. Accessed in September 2015.

WRI. (2005). http://www.wri.org/our-work. Accessed in September 2015.
Young, M. D., & Cabezas, H. (1999). Design sustainable processes with simulation: The waste reduction (WAR) algorithm. *Computers & Chemical Engineering, 23*, 1477–2491.
Young, D., Scharp, R., & Cabezas, H. (2000). The waste reduction (WAR) algorithm: Environmental impacts, energy consumption and engineering economics. *Wast Management, 20*, 605–615.